GROWING GATWICK

Vincent J. Bunce

Series Editors:
Michael Naish and Sue Warn

Acknowledgements

I am grateful to the following people for their help in providing information and support in the preparation of this book: E M Holdsworth, Secretary of the Gatwick Airport Consultative Committee; Ian Ockenden for assistance with the Horsham Oak Tree Survey; and Richard Wainwright of the Research Department of the British Airports Authority.

I would also like to thank Chronis, Kirsten, Andrew, John, Leonora, Dominic, Barry and Steven who were, unknowingly, the inspiration behind this unit and who helped trial the early drafts of this book.

The publishers are grateful to the following for permission to reproduce copyright material;
the author, Robert M Eadie for his letter from *Kent & Sussex Courier* 24.7.87; Ewan MacNaughton Associates for an extract from the article 'Village fears over increased pressure for new homes' by A J McIlroy from *Daily Telegraph* 17.8.87, © The Daily Telegraph plc; Newspaper Publishing plc for an extract from the article 'Consortium plans London airport in Thames estuary' by David Black from *The Independent* 9.3.90, the article 'Anger at second Gatwick runway' by Nicholas Schoon from *The Independent* 22.3.89 and the article 'Forlorn battle base favourite to expand' from *The Independent* 27.2.89; the Chairman, Brendon Sewill CBE on behalf of the Gatwick Area Conservation Campaign for an extract from *GACC Newsletter* No 26, July 1986; Surrey & South London Newspapers Ltd for the article 'Jet crash alert' from *Crawley News* 29.7.87; Sussex Newspapers for the article 'County will resist night fly pressure' from *Crawley Observer* 29.7.87.

The publishers are grateful to the following for permission to reproduce photographs, graphics and tables and maps: Ace Photo Agency, fig. 4a, 38; Adams Picture Library, fig. 31; J Allan Cash Ltd, fig. 5; *Crawley Observer*, fig. 29, 36; Departments of the Environment and Transport, fig. 32; Financial Times graphic: 'Plan for Thames Estuary Airport', fig. 47; Gatwick Airport Ltd, figs 7, 9, 10, 11, 14, 16, 17, 18, 19, 20, 22, 24, 25, 26, 33, 37; Richard Gardner fig. 1b; Controller of Her Majesty's Stationary Office, table 'Future of UK Air Traffic Growth' December 1988, fig. 23; Horsham Friends of the Earth Oak Tree Survey, fig. 39; The Hulton Picture Company, fig. 45; *The Independent* graphic 'Estimated Increase in Traffic', 27.2.89, fig. 42; Adrian Meredith, Solo Syndication, cartoon by Jak, *London Evening Standard*, fig. 2; 'How Air Travel Has Taken Off', *The Daily Mail* 25.6.88 Photo: John Frost Historical Newspaper Service, fig. 3; Topham Picture Source, fig. 40.
We were unable to trace the copyright holder of fig. 6 and would be grateful to receive any information that would enable us to do so.

Longman Group UK Limited
Longman House, Burnt Mill, Harlow, Essex, CM20 2JE, England and Associated Companies throughout the World.

First published 1994
ISBN 0 582 075580

Set in 10/12pt Times (Linotron)
Produced by Longman Singapore Publishers Pte Ltd
Printed in Singapore

Cover illustration by Stuart Briers

Contents

Airports: Gatwick

- 18.80 million passengers used Gatwick in 1991/92
- over 100 airlines use the airport, flying to 134 destinations
- there are 20 hotels within easy reach of the airport with a total of over 2000 rooms
- services at the airport include shops and banks, restaurants and bars and even a chapel

THE POST HOUSE HOTEL

Gatwick Airport

With summer now under way we are seeking the following personnel:

ROOM MAIDS (Full time)

To assist in the servicing and cleaning of guest bedrooms. The position involves working from 7.00 am to 3.00 pm five days per week. Experience is not necessary as full training is provided.

BREAKFAST CHEF

WORLDWIDE AIR CARGO

EXPORT CLERK

A vacancy has arisen in our Export Dept. Applicants should have a minimum of 2 years experience with general knowledge of airfreight and ACP80. Full driving licence required

Phone Paul Coomber, Crawley 5148 for application form

COPY TYPIST/ ADMINISTRATION ASSISTANT

Our busy Cabin Staff Department based at Gatwick Airport currently has a vacancy for an accurate typist.

The successful applicant will have the ability to cope under pressure, be of smart appearance, have a good telephone manner and a sense of humour.

Anger at second Gatwick runway

By Nicholas Schoon

The Select Committee's unequivocal recommendation for a second runway at Gatwick brought an instant and angry rejection from councils and MPs.

The Government will find it a highly unpalatable suggestion. The world's second busiest international airport is surrounded by a Tory-voting heartland, and any suggestion that the proposal should be taken up will create enormous anger.

Building the new runway in order to meet a doubling in flights over the next 16 years would require Parliament to set aside a legal agreement made between the BAA and West Sussex County Council in 1979 that no application would be made to build a second runway for the next 40 years.

A new runway, parallel to the existing one, would be the simplest solution but it would cause further massive disruption in a once rural area which is already struggling with one of the highest growth rates in the country.

Peter Bryant, director of planning for West Sussex, said: 'The MPs are flying in the face of reality, and I think they're being wholly irresponsible. They can't have examined the practical issues so their comments must be of limited value.' Michael Sanders, chief executive of Crawley Borough Council which includes the airport, said: 'We're appalled. We've always been supportive of Gatwick, but we could not cope with the extra growth a second runway would cause.'

Source: *The Independent* 22 March 1989

  Gatwick

'The world's major airports are cities in themselves, designed to serve the needs of travellers, staff, the "meeters and greeters", and those who come to absorb the atmosphere and see the sights of modern air transport. They are interchange points between air services and every form of surface transport. They are also, by their very proximity, closely involved with the local communities and immediate environs.'

(Source: *The Flier's Handbook* edited by Helen Varley, 1978)

- 500 trains call at the airport's British Rail station each day
- there were 163 000 flight movements (arrivals and departures) at Gatwick during 1991/92
- 21 807 people are employed at Gatwick Airport in some capacity
- the opening of Gatwick's new North Terminal in March 1988 brought the capacity up to 25 million passengers a year

PARLIAMENTARY QUESTIONS

Aircraft Noise

The Hon. Nicholas Soames MP (Conservative - Crawley) asked the Secretary of State for Transport to make a statement on aircraft noise at Gatwick airport. Mr Michael Spicer, Parliamentary Under Secretary of State for Transport replied: 'The noise climate around Gatwick has improved significantly over the last 10 years. Air transport movements have increased by 91 per cent over this period but the area in which people generally experience annoyance has decreased by 29 per cent.'

(Hansard, 18 March 1985, Col 827)

Moving away is just plane sense

May I politely suggest that the only solution to Mr Burren's problem of aircraft noise (July 10) is for him to move to either the Yorkshire Dales, or the Scottish Highlands, in spite of his obvious liking for the Tunbridge Wells area.

We have two major airports which have to be within easy reach of London, and it is obvious that aircraft flight over many built-up areas is unavoidable, and has to be accepted. Whether or not this is 'progress' must be a matter of opinion.

Here, at Mark Beech, we too are on a direct flight path to Gatwick, and on occasions, get a procession of planes at about 3 minute intervals, and flying lower than at Rusthall, I would think. I have often counted 12 in 40 minutes some evenings. On the other hand many days and nights often go by without us hearing a plane. Flights do get 'staggered' to a certain extent, by reason of different approaches to the runway, according to wind direction.

We, also, appreciate peace and quiet, and the benefits of living in this lovely countryside, but have never felt the need to move because of aircraft noise. We are grateful not to be living at Horley or Copthorne, where, indeed, one might have some cause for concern.

Finally, I believe that the airlines pay vast sums to use the airports - so why another £1 penalty for the privilege of flying over Tunbridge Wells?

(Mr) R. M. Eadie

Source: *Kent and Sussex Courier*, 24 July 1987

Jet crash alert

Police closed road and rail links at Gatwick on Thursday when a jet looked as if it might crash land.

The plane - with 267 passengers on board - had taken off for Cincinatti when the alert was sounded.

Pilot Jess Bottonhoff reported the wing flaps on the Delta Airlines Lockhead Tristar had jammed on take-off and the jet lacked full braking power.

Police sealed off the A23 road and the London-Brighton railway line as the plane made an emergency landing.

Live railway lines were shut down to minimise the risk of an explosion and ambulance and fire crews were alerted.

The plane dumped fuel in the Channel before landing safely.

Delta is now under investigation by the US Federal Aviation Administration after SIX incidents in only TEN days.

Source: *Crawley News*, 29 July 1987

Introduction

This book is concerned with how improvements in communications, or more specifically the movement of people and goods by air, affect the environment. By focusing on air transport, and taking as a case-study the development of Gatwick Airport in West Sussex, some of the implications of transport development for people and environments are considered. The book is appropriate for students studying the 'A' or 'AS' option modules on Mobility and the Environment and The Communications Revolution in the ULEAC Geography 16–19 syllabuses.

People are travelling much more frequently than in the past, for pleasure as well as for business reasons. This increase in travel has been facilitated by improvements in communications. The development of the motorway network, electrification of the railways and gradual expansion of the air-route network in the UK have all contributed to increasing the frequency of passenger journeys. All these improvements in transport technology have particular effects on the way space is used.

Among the most important means of communication today is air travel. Air-route networks have become extensive in many areas. The nodal points of these networks, that is the airports, are where the impacts of this particular form of transport on people and their environments are maximised. The airport is the focus of a very real people-environment conflict. The growth of air travel has resulted in some 'gainers' and some 'losers', and this book will help to identify them, and to consider the planning and environmental problems that airport development embodies.

The book begins with a general introduction to the function and form of airports. In Exercises 1 and 2, the way in which people perceive airports is explored, then some of the reasons behind the massive recent expansion in air transport are considered in global and domestic frameworks. The detailed case-study of Britain's second largest airport at Gatwick near Crawley in West Sussex begins in Exercise 3. A 'General Model of Airport Location' is considered, then attention is given to the location and role of Gatwick. A variety of simple field techniques are suggested. These can be used at any airport to gain a deeper insight into its function, layout and dynamics.

The principal task that transport planners have is to forecast future demand. This is essential because the construction of new transport infrastructure like motorways, railway networks and airports has a long 'lead-time'. A new airport runway or terminal may take 7 to 10 years to commission from the date that permission to build is given. Exercise 4 considers some of the techniques used to forecast air traffic growth, and the difficulties associated with them. It applies them to the situation at Gatwick in particular, as well as to the whole London area. The rationale for expanding the facilities at Gatwick is then clarified in terms of expanding demand, anticipated future growth, positive effects on the local economy and employment prospects, so that it may be investigated.

A selection of evidence relating to the different impacts the airport has had is put forward in Exercise 5. Noise, pollution and other environmental manifestations of the growth of the airport are discussed, and students have an opportunity to consider much of the research that has been undertaken into the impact of Gatwick Airport. Studies done by the government's Warren Spring Laboratory are examined along with evidence from the official noise monitoring stations around the airport and some collected by local Friends of the Earth campaigners.

The book concludes with a role-play exercise which focuses on the very real and important 'issue' of whether a second runway should be constructed at Gatwick, as was suggested in early 1989 by the Transport Committee of the House of Commons. This will enable the key questions and evidence contained in the book to be re-examined in the context of a simulated public inquiry.

The development of airports, whether it is a new terminal, more night flights or additional runway capacity, is an extremely emotive topic. It is a people–environment issue laden with value judgements about what is best and for whom, and complicated by difficulties in forecasting trends and quantifying benefits and impacts. This book presents some of the evidence which relates to one real conflict at Gatwick Airport. The arguments surrounding this issue are being rehearsed now, and seem likely to resurface some time in the future.

EXERCISE 1

Why do we need airports?

For most of us, airports are places where we spend as little time as possible, usually at the start and end of our holiday. But holidaymakers are not the only people who 'experience' airports. Business people, journalists and long-term migrants also often travel by air.

People who live near airports also notice their presence in various ways of course. For many communities and the people who live in them, airports provide a major source of revenue and employment. Some places owe their whole existence to proximity to an airport. Such proximity also brings its fair share of problems – including noise, air pollution and congestion.

What image do you have of airports? What are they like? Brainstorm up to 10 words which you associate with airports. Compare your list with others in the group to build up a picture of the main features of a large international airport.

As the growth in air traffic continues almost relentlessly, more and more people are using airports and experiencing the benefits and problems which they bring. Any discussion about airports should raise many key questions and issues: Why do we need airports? Why are they growing so rapidly? And what are the impacts and consequences of airports?

Figure 1 provides you with some information about Gatwick Airport in Sussex. This is Britain's second largest airport. Study the material carefully and make two lists: one of the advantages the airport has brought to its local area, and a second of the disadvantages which are associated with its development. Extend your list by comparing it with those of others in the group.

For passengers airports are, by definition, a 'means to an end'. They are a necessary evil which most people may only encounter perhaps twice each year. However, in recent summers airports all over Europe have attracted considerable media attention. This has been due firstly to terrorist violence and the increased security needs that this has brought, and secondly to delays and congestion (see Figure 2) caused partly by the general increase in air traffic and partly as a reaction to strike action by air-traffic controllers.

Even if our acquaintance with airports is occasional and brief, it is important to remember the key role they play in our modern world. Large modern airports constitute major communities in their own right, and are strategic nodes in the global transport network. They are responsible for transporting huge quantities of cargo as well as large numbers of people.

Within the communities where they are located, airports need to be integrated into local networks. The flows of people, traffic and cargo around an airport need to be managed just as carefully as take-offs and landings at the airport itself. Local councils bear the responsibility for coping with the planning implications of airports located in their areas.

In small groups, consider how a major international airport like Gatwick makes demands on the local county council. What special provisions are required? Which aspects of a council's normal planning provision are likely to be affected by a large airport?

Local councils often have very little control over the initial location of an airport. However, they must cope with the demands subsequently made by any airport constructed in their area. A major airport will affect the lives of thousands of people either directly or indirectly. It may be a political and environmental liability as well as an economic benefit.

The extent of an airport's impact on people ultimately depends on the importance given to environmental considerations in managing the airport's growth and development.

'You have it guaranteed for two weeks every year. Just think of the money you save.'

Figure 2

EXERCISE 2

The expansion of air transport

Since Orville and Wilbur Wright first flew at Kitty Hawk in North Carolina, USA at the start of the century, an estimated 15 billion people have flown. This figure represents the equivalent of three flights on average for every person alive in the world today. It is also a measure of the spread and popularity of air travel as a means of mass transport.

Why do you think air travel is so popular? What advantages can it offer compared with other modes of transport? Are there any major disadvantages associated with travelling by air?

The industry now extends into every country in the world, and has brought about a transport network that is truly global in character.

The growth of global air traffic

Around one billion people were carried on flights worldwide in 1990. This is more than ever before. Figure 3 shows how the growth in air traffic has continued unabated since the start of the century, slowly at first and more rapidly in recent decades. Most projections show this trend continuing for many years to come.

The most important decade for air transport was the 1960s. Until then, travelling by air was the preserve of quite a small group of affluent people. It remained a relatively exclusive means of transport until the mid-1960s, when economic growth and the higher living standards and expectations it brought, together with improvements in technology, combined to increase the speed and reduce the price of travel by air relative to other forms of transport. The prospect of holidays abroad for large numbers of people thus became a reality. This laid the foundation for the expansion in air transport throughout the 1970s and 1980s.

The development of larger planes helped to reduce the cost of flying. Increased seating capacities of wide-bodied jets like the Boeing 747 'jumbo jet' compared with earlier models (see Figure 4) meant that larger numbers of people could be carried, and thus that unit costs could be further reduced. This brought medium-haul flights, covering a range of European destinations, at least within the reach of a mass market.

In groups of three or four, compile a list of reasons which might explain the increase in air traffic that has occurred in recent decades. Compare your list with those produced by other groups, and try to produce, by consensus, an agreed list of the five principal reasons for the increase. Don't forget to consider users of air transport other than passenger traffic.

It was in the mid- to late 1960s that the concept of 'package' holidays to overseas destinations became popular. Typically these packages included accommodation for one or two weeks in a hotel together with return flights from a British airport. They were relatively inexpensive because the tour-operators could pre-book large numbers of hotel rooms, thus guaranteeing in advance that they would be full, and also charter airplanes so that they too were used to maximum efficiency.

This sort of mass tourism has been likened to farming, with the tourists themselves being herded around 'like cattle' in large groups, in the name of economy. Destinations in southern Spain, France and Portugal became popular. High temperatures and sunny days were guaranteed. These were important advantages for visitors from northern Europe. A massive expansion in hotel construction took place (see Figure 5), and the tourism boom spread around the Mediterranean, then to the Greek islands and more recently to North Africa. The tourists were not only Britons, but came from all over Europe. The tourist industry is now so large that it accounts for one out of every sixteen jobs worldwide.

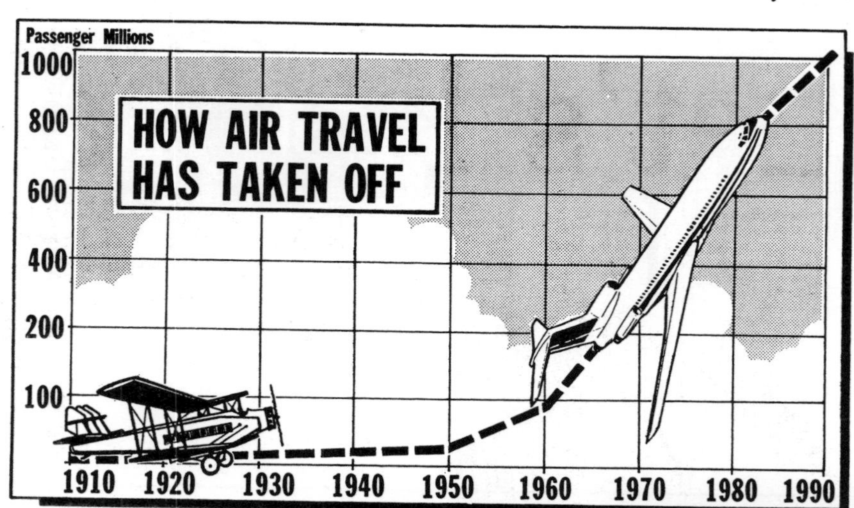

Figure 3 Global increase in air travel from 1910

Figure 4a *Changes in aircraft size and capacity*

Figure 4b *Growth in wide-bodied aircraft at Gatwick*

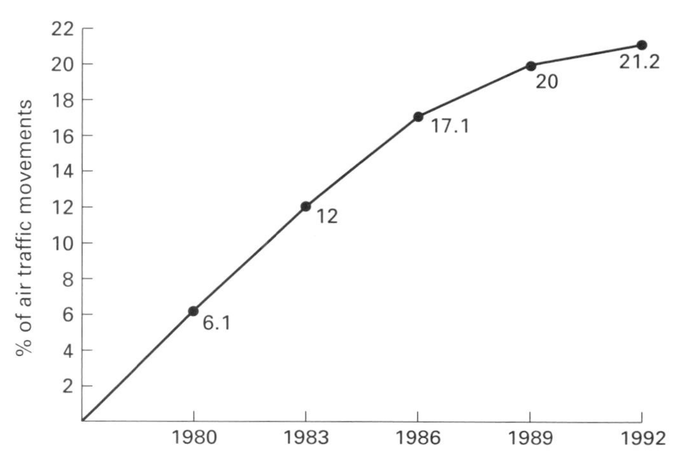

Some countries are heavily reliant on it. Spain, for example, derives 10% of its Gross National Product from tourism, which also employs 10% of its active workforce.

While European destinations attracted substantial numbers of people, the tourism boom was not confined to this continent. Indeed, the 1970s and 1980s saw a rise in the number of tourists making long-haul journeys. North America in particular became a major destination for visitors, with Florida an important tourist focus. Many other locations are now attracting record visitors, including the Caribbean, Brazil and Mexico, as well as other parts of the world like South-East Asia and Africa.

Another reason for the growth in air traffic generally, and tourism in particular, was the increased competitiveness of the airline industry. Several entrepreneurs started businesses carrying passengers on routes that they felt were profitable. Companies like Laker Airways and People Express operated successfully for a while on the prestigious transatlantic route before being forced out of business. A more recent arrival is Richard Branson's Virgin Atlantic Airways Ltd. On short- and medium-haul routes, airlines specialising in charter operations like Dan-Air, Britannia and Monarch increased competition and reduced seat prices.

Figure 5 *Hotel development in the Mediterranean*

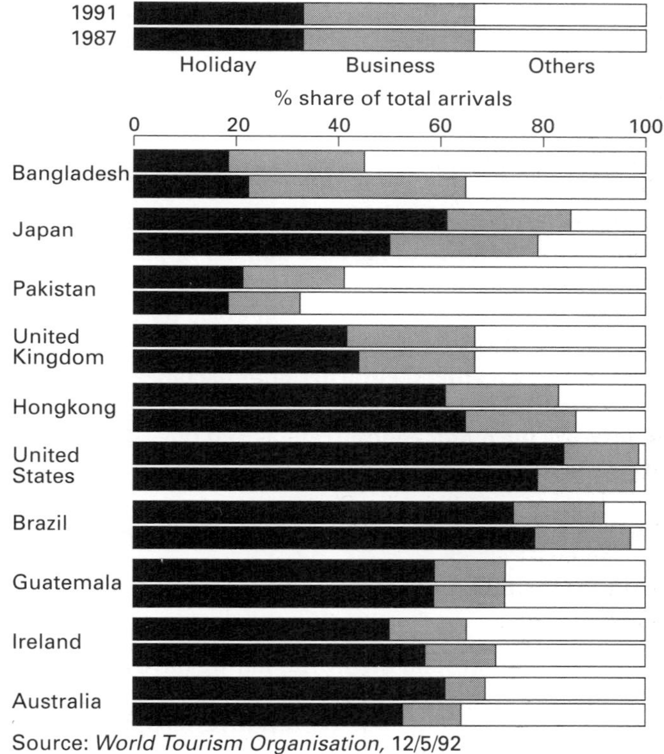

Source: *World Tourism Organisation,* 12/5/92

Figure 6 *Purpose of travel in 1991 and 1987*

Going on holiday is not the only reason people travel overseas. Figure 6 shows the purposes of travel recorded amongst arrivals in 10 countries. Summarise what it shows, and explain the main trends between 1987 and 1991. Can you think of any major 'purposes' of air travel other than those stated?

a World airport rankings 1991

Top Twenty	AIRPORT	(Millions of Passengers)
1	Chicago O'Hare	59.9
2	Dallas Fort Worth	48.2
3	Los Angeles (International)	45.7
4	Tokyo (Haneda)	42.0
5	**London – Heathrow**	**40.2**
6	Atlanta (Hartsfield)	38.9
7	San Francisco	31.2
8	Denver (Stapleton)	28.3
9	Frankfurt	27.4
10	Miami	26.6
11	New York (Kennedy)	26.3
12	Osaka	23.5
13	Paris (Orly)	23.2
14	New York (Newark)	22.3
15	Phoenix	22.1
16	Paris (CDG)	21.6
17	Boston (Logan)	21.5
18	Honolulu	21.2
19	Detroit	20.7
20	Minneapolis	20.6

Gatwick (1991) = 18.69m

b World airports : International passengers 1991

	Rank	International Passengers 000's	Terminal Passengers 000's	% of Total
Heathrow > >	**1**	**33,519**	**40,248**	**83.3**
Frankfurt	2	20,972	27,369	76.6
Paris (CDG)	3	19,389	21,612	89.7
Hong Kong	4	19,158	19,158	100.0
Tokyo (Narita)	5	17,743	20,622	86.0
Gatwick > >	**6**	**17,677**	**18,609**	**95.0**
Amsterdam	7	16,082	16,183	99.4
Singapore	8	15,065	15,065	100.0
New York (JFK)	9	14,668	26,335	55.7
Zurich	10	11,185	11,835	94.5

Updated > 3.6.92

Source : BAA plc

Figure 7 The world's largest airports (measured by number of passengers) in 1991

Overall, in 1987 tourists accounted for 55% of foreign travel and business people for 21%. Both these figures were slightly higher than in 1980, and both hide massive absolute increases in numbers travelling. Most foreign travel is by air because it has become easier, quicker and cheaper to travel in this way, as the global network of air transport has grown.

Figure 7 shows the world's 20 largest airports in 1991 as measured in terms of the total number of passengers, and the 10 largest international airports.
a) How do Britain's airports fare in world terms?
b) Plot the 20 largest airports on a world map, devising a method of indicating their order of importance.
c) Comment on the spatial distribution of large airports. Where are the major concentrations and the void areas? How can these be explained?

What is happening to air travel inside the UK?

Air travel in the UK has expanded in recent years, in line with the trends around the world that we have already considered. There are 36 principal airports (see Figure 8). Three of the top five airports, including Heathrow and Gatwick, are owned and operated by the British Airports Authority (BAA). Most of the remainder are public limited companies (PLCs) which are owned by their local authorities.

Airport capacity (passengers and cargo) is clearly concentrated around London, with Heathrow and Gatwick the principal 'gateway' airports. Manchester follows close behind, but will be overtaken as the capacity of Stansted, designated as London's third airport, increases to meet demand. A new £400 million terminal building was opened in March 1991 to help meet this increase. The concentration of airport capacity in London and the south-east has been recognised by the government. Under the 1986 Airports Act, 12 airports became PLCs and were removed from the direct control of their local authorities in an attempt to promote the growth of regional airports.

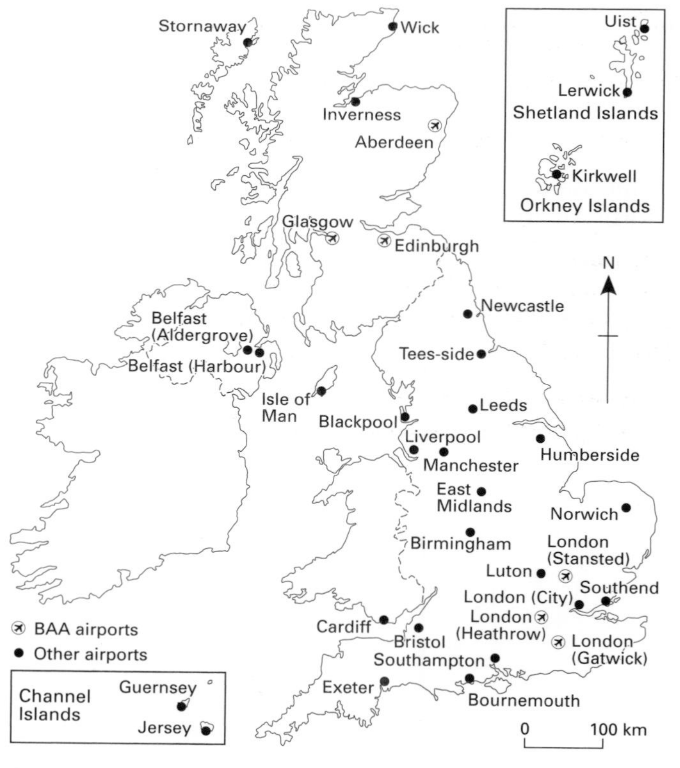

Figure 8 The UK's airports

Airport	TERMINAL PASSENGERS 000's	% change	AIR TRANSPORT MOVEMENTS 000's	% change	CARGO (TONNES) 000's	% change
ABERDEEN	**2,014.6**	**3.6**	**85.5**	**2.8**	**5.8**	**−8.4**
Belfast	2,169.2	−5.5	43.5	−2.7	24.0	1.3
Belfast City	536.2	−1.9	18.4	−0.1	0.8	−53.8
Birmingham	3,251.2	−6.8	66.1	0.2	26.0	21.2
Blackpool	81.7	−17.6	7.6	−8.4	1.0	−20.8
Bournemouth	201.0	−4.3	5.1	−29.7	8.2	−1.2
Bristol	782.8	1.1	22.0	32.6	0.1	−88.5
Cardiff–Wales	518.2	−12.9	12.3	13.2	0.5	–
East Midlands	1,142.7	−10.8	30.3	−6.3	8.4	−24.7
EDINBURGH	**2,342.8**	**−6.1**	**49.7**	**3.7**	**1.1**	**−4.7**
Exeter	164.1	−24.5	6.9	−31.2	neg	–
GLASGOW	**4,153.8**	**−3.1**	**72.0**	**1.5**	**14.9**	**−20.8**
Guernsey	748.2	−13.3	39.6	−15.7	5.4	−11.6
Humberside	108.9	−26.6	11.6	−18.5	neg	–
Inverness +	210.2	−0.4	6.0	−6.6	1.3	−0.1
Isle of Man	492.1	−11.3	14.6	−6.4	2.7	−17.4
Isles of Scilly (St Mary's)	101.7	−11.6	8.4	−20.6	0.7	40.6
Jersey	1,637.5	−12.3	46.6	−17.7	3.3	−14.0
Kirkwall	102.4	−3.1	10.9	−7.7	0.4	−3.1
Leeds/Bradford	638.9	−23.4	18.9	−18.8	0.4	−35.1
Lerwick	10.7	−16.2	2.8	−11.0	neg	–
Liverpool	465.7	−7.4	21.7	−8.4	26.0	−1.1
London City Airport	171.9	−25.3	9.7	−27.0	neg	–
LONDON (GATWICK)	**18,690.2**	**−11.2**	**162.8**	**−13.5**	**202.7**	**−7.8**
LONDON (HEATHROW)	**40,248.4**	**−5.6**	**361.1**	**−1.7**	**661.1**	**−5.3**
LONDON (STANSTED)	**1,683.6**	**45.9**	**35.3**	**48.3**	**34.6**	**5.2**
Luton	1,941.5	−27.5	36.6	−22.2	32.7	2.2
Manchester	10,150.0	−0.1	126.5	3.2	66.2	−9.1
Newcastle	1,542.8	−1.9	29.2	5.0	0.9	16.8
Norwich	176.7	−14.5	9.3	−3.4	0.2	1.1
Penzance (Heliport)	96.8	0.0	5.2	7.0	0.4	−19.0
Prestwick	35.6	−63.1	3.6	−9.9	15.3	−1.9
SOUTHAMPTON	**425.7**	**−13.4**	**19.7**	**−2.3**	**1.0**	**−11.8**
Southend	20.8	−82.5	6.5	−39.3	13.5	5.4
Stornoway	80.2	−3.9	3.8	−3.6	0.5	−1.3
Sumburgh	432.6	0.1	21.1	−13.5	1.0	−9.6
Tees-side	303.6	12.6	11.4	7.8	neg	–
Uist	85.2	3.2	4.5	−2.2	0.2	−8.8

+ = Financial Year 1991/92

Sources: BAA plc, Civil Aviation Authority and individual airports.

Figure 9 Air traffic statistics – UK Airports 1991/92

What conclusions can you draw about the health of the UK's airports from the statistics in Figure 9? Why is any conclusion based solely on the data provided necessarily a weak one? What other data would help you form a better judgement?

Devise a single index to show the relative growth of UK airports, then map the index on an outline map of the UK.

There have been increases at most airports in the number of passengers, the volume of air movements and the amount of cargo handled. In part, the increase in passenger traffic, especially at regional airports, is a result of the increased use of air travel by the business community, for whom convenience and speed are key factors. Several carriers now have well-established and scheduled 'shuttle' services between the principal UK cities. New airlines, like British Midland and Air UK, have entered the market on these routes to make the cost of a seat much more competitive.

Passengers (000's) – 10 year record

	1982/83	1983/84	1984/85	1985/86	1986/87	1987/88	1988/89	1989/90	1990/91	**1991/92**
Heathrow										
Total Terminal Passengers	26,277.8	26,976.5	29,866.2	31,421.4	31,713.2	35,638.5	38,058.1	40,315.9	41,217.2	**42,045.1**
% change	−0.7	2.7	10.7	5.2	0.9	12.4	6.8	5.9	2.2	**2.0**
– Domestic	4,210.2	4.503.7	5,191.2	5,472.1	5,661.2	6,381.4	6,956.0	7,254.1	7,105.6	**6,812.5**
– International	22,067.6	22,472.8	24,675.0	25,949.3	26,052.0	29,257.1	31,102.1	33,061.8	34,111.6	**35,232.6**
– *Non-scheduled**										**85.0**
Transit	330.6	313.5	306.3	323.2	379.0	334.1	322.9	296.1	290.8	**258.1**
Gatwick										
Total Terminal Passengers	11,288.9	12,745.3	14,228.0	15,211.7	16,596.1	20,091.0	21,056.6	21,206.4	20,421.7	**18,874.0**
% change	3.7	12.9	11.6	6.9	9.1	21.1	4.8	0.7	−3.7	**−7.6**
– Scheduled	4,582.2	5,034.8	5,832.3	6,699.6	6,993.9	8,807.1	9,778.9	11,450.0	11,993.2	**10,505.6**
– Non-scheduled	6,706.7	7,710.5	8,395.7	8,512.2	9,602.2	11,283.9	11,277.7	9,756.5	8,428.5	**8,368.4**
Transit	204.3	163.7	214.0	196.1	155.1	199.5	102.1	105.9	144.1	**123.8**
Stansted										
Total Terminal Passengers	298.6	357.2	546.8	504.4	555.9	722.6	1,124.6	1,314.3	1,127.5	**1,918.0**
% change	9.2	19.6	53.1	−7.8	10.2	29.9	55.6	16.9	−14.2	**70.1**
– Scheduled	29.6	40.6	105.5	121.4	141.6	237.8	288.9	352.1	410.4	**1,240.7**
– Non-scheduled	269.0	316.5	441.3	383.0	414.4	484.8	835.7	962.2	717.1	**677.3**
Transit	8.6	10.5	12.4	23.8	21.0	35.4	64.6	52.7	22.1	**37.3**

*Note: Until July 1991 Heathrow's traffic was almost entirely scheduled. Since that date, non-scheduled services have been allowed to operate at Heathrow.

Cargo (tonnes) – 10 year record

	1982/83	1983/84	1984/85	1985/86	1986/87	1987/88	1988/89	1989/90	1990/91	**1991/92**
Heathrow										
Total Cargo	445,412	487,402	544,027	528,775	545,718	594,106	656,107	699,878	671,900	**691,371**
% change	0.6	9.4	11.6	−2.8	3.2	8.8	10.4	6.7	−4.0	**2.9**
Scheduled	442,969	486,408	543,702	527,448	544,327	592,187	653,603	696,670	671,757	**691,189**
– Non-scheduled	2,443	994	325	1,327	1,391	1,919	2,504	3,208	143	**182**
Total Cargo & Mail	505,522	552,713	615,814	602,636	617,975	662,292	723,015	774,873	751,988	**768,969**
Gatwick										
Total Cargo	114,295	115,643	146,262	158,286	168,522	199,555	193,641	212,756	219,616	**198,341**
% change	−13.6	1.2	26.5	8.2	6.5	18.4	−3.0	9.8	3.2	**−9.7**
Scheduled	85,657	100,565	111,541	130,743	145,175	176,881	176,087	191,961	200,844	**176,998**
– Non-scheduled	28,638	15,078	34,721	27,543	23,347	22,674	17,554	20,795	18,772	**21,343**
Total Cargo & Mail	124,822	127,590	156,561	170,928	182,005	215,871	205,384	220,240	228,028	**208,589**
Stansted										
Total Cargo	7,721	18,481	14,599	10,695	13,993	18,288	27,671	31,561	32,799	**38,973**
% change	20.5	139.4	−21.0	−26.7	30.8	30.6	51.3	14.0	3.9	**18.8**
Scheduled	1,379	2,812	4,570	2,984	4,967	7,662	17,045	14,236	19,736	**27,478**
– Non-scheduled	6,342	15,669	10,029	7,711	9,026	10,626	10,626	17,325	13,063	**11,495**
Total Cargo & Mail	10,372	20,195	15,953	12,432	16,016	20,025	32,122	32,535	33,595	**39,889**

Note: Since 1988/89 Gatwick cargo tonnage has been under-stated as a result of incomplete reporting to Gatwick Airport Limited.

Aircraft movements (000's) – 10 year record

	1982/83	1983/84	1984/85	1985/86	1986/87	1987/88	1988/89	1989/90	1990/91	**1991/92**
Heathrow										
Total Air Transport Movements	252.8	263.3	274.2	285.7	292.3	310.4	330.4	351.3	361.2	**373.9**
% change	2.9	4.1	4.1	4.2	2.3	6.2	6.4	6.3	2.8	**3.5**
General Aviation	18.7	20.6	20.5	21.2	21.4	20.5	18.1	17.2	15.7	**15.8**
Other	3.4	3.0	4.1	5.2	5.0	5.9	6.6	6.5	6.6	**4.5**
Total Aircraft Movements	274.9	286.9	298.8	312.2	318.8	336.8	355.1	375.0	383.5	**394.2**
Gatwick										
Total Air Transport Movements	132.6	136.0	141.0	151.0	155.2	177.1	182.7	190.9	181.9	**163.2**
% change	4.5	2.5	3.7	7.1	2.8	14.1	3.2	4.5	−4.7	**−10.3**
General Aviation	12.3	10.9	10.0	10.2	9.4	8.4	7.8	6.8	5.1	**4.3**
Other	6.4	6.0	8.8	9.6	9.2	9.7	9.7	10.1	9.4	**6.9**
Total Aircraft Movements	151.2	152.8	159.8	170.8	173.8	195.2	200.2	207.8	196.4	**174.4**
Stansted										
Total Air Transport Movements	7.7	8.6	12.8	14.2	16.8	20.5	24.7	25.1	22.7	**41.4**
% change	25.6	11.1	48.6	10.8	18.3	22.4	20.6	1.6	−9.5	**82.4**
General Aviation	14.1	15.4	14.9	14.7	13.8	13.1	13.1	13.3	12.7	**10.4**
Other	11.5	10.9	12.7	12.4	16.2	13.1	11.7	11.7	8.9	**6.8**
Total Aircraft Movements	33.4	34.9	40.4	41.2	46.8	46.7	49.5	50.1	44.3	**58.6**

Source: BAA plc

Figure 10 London's three airports compared

Using Figure 10, draw graphs to show how London's three main airports compare with respect to:
a) total passengers handled 1982/83–1991/92
b) total cargo and mail handled 1982/83–1991/92
c) number of aircraft movements 1982/83–1991/92
Write a brief commentary on the trends you have shown graphically. Refer to percentage changes between 1982 and 1991, which you will need to calculate.

The future?

An enormous amount of time and money is being spent trying to estimate future changes in the volume of air traffic. There are, of course, sound pragmatic reasons for this expenditure, but these do not make it any easier to predict more accurately what will happen.

From the trends on your graphs, what do you think will happen in the next 10 years to the volume of air movements and the number of passengers? Can you think of factors which may reduce the sustained pattern of growth that there has been to date?

For what reasons are estimates made of the likely future volume of air traffic? List the various groups, organisations and individuals likely to be involved in discussion and debate about future provision of airport capacity in Britain.

If, as seems likely, demand for increased airport capacity continues to rise, then it is in London and the south-east of England that the debate about how to meet this demand will be the most vigorous. Here the potential future demand is greatest, and the consequent pressure on land and the environment the most intense.

In the rest of this book we are going to examine Britain's second largest airport, Gatwick in Sussex. What sort of airport is it? How quickly is it growing, and will present trends continue? What impact does the presence of an airport have on local people and the environment? These same questions could be asked about your local airport.

EXERCISE 3

What sort of airport is Gatwick?

Although all airports exist as arrival and departure points for people and goods in transit, there are important differences in their precise function and characteristics, as well as in their comparative sizes. The relative importance of cargo and passengers varies from place to place, as does the percentage of domestic and international passengers and the types of aircraft which use the airport. Airport locations, however, tend to have certain common characteristics.

Where is Gatwick Airport?

Gatwick Airport is located 45 kilometres south of London in the Sussex countryside, close to Crawley new town. Several transport links exist that enable the airport to be reached easily and quickly by private car as well as public transport. Links are good in all directions, but are particularly so with central London (see Figure 11). There is a rail link to London Victoria station with a journey time of only 30 minutes, and to London Bridge station. There are also coaches into London as well as to London's other main airports. Public transport is generally quite good.

Road users can reach Gatwick via the airport's direct link to the M23 motorway, from which they have access to the M25 orbital motorway around the capital. The M25 westbound provides access to the M3, M4 and M40 motorways, and to London's other principal airport at Heathrow. Eastbound, the M25 links Gatwick with the M2 and M20 motorways into Kent and the Dartford Tunnel as well as with the Channel ports.

Does Gatwick's location have anything in common with other airports?

Make a list of the general site and situation requirements for a modern international airport. Use Figure 11 to help you, together with any knowledge you have about other large airports.

Airports which are capable of taking the passenger aircraft used today need large flat sites. Weather conditions must, on balance, be moderate with good visibility for a large proportion of the year. Fog should not be a common phenomenon. For obvious safety reasons, airports are not often situated in the middle of high-density concentrations of population. However, they need to be close enough to cities to allow easy and quick access for passengers travelling to and from the airport. Good links to national transport networks are also a key factor.

Figure 11 Location of Gatwick airport

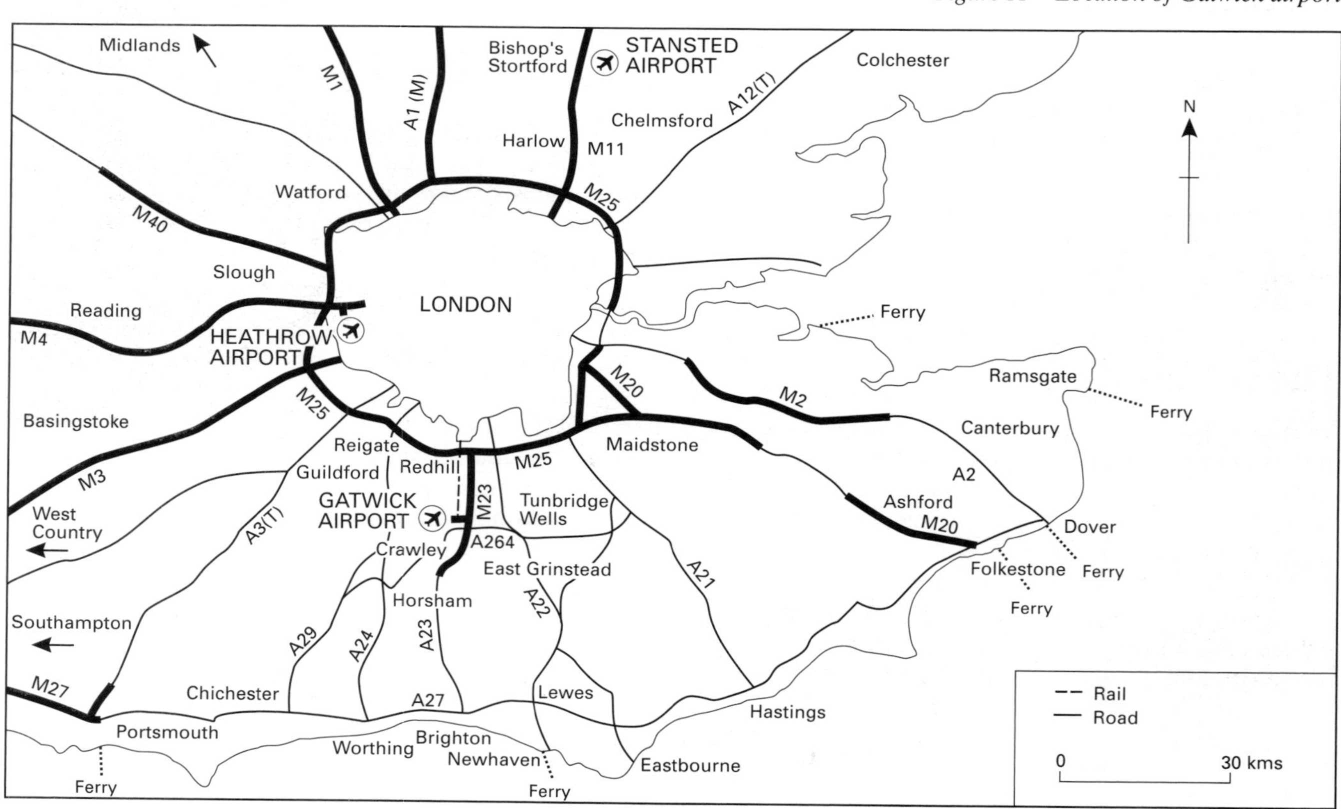

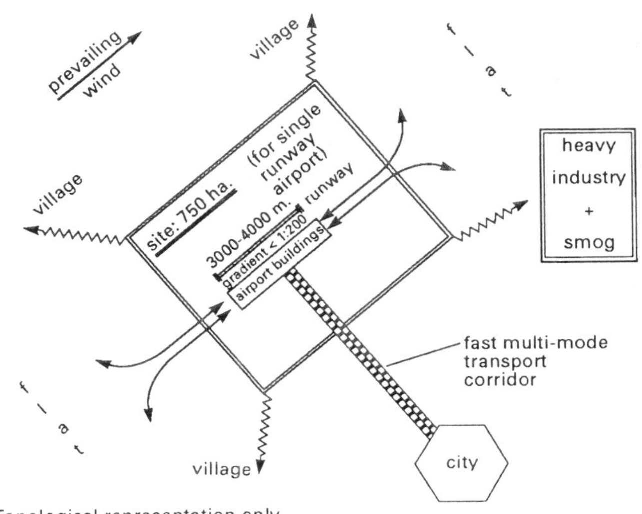

Topological representation only

⟿ = undefined but large distance

Figure 12   General model of airport location

All these requirements can be incorporated into a 'General Model of Airport Location' (see Figure 12), which can then be used to evaluate the sites of individual airports. When you have looked at Figure 12, try these questions:

a) What are the three most important criteria for a good airport location? Does Gatwick meet these three criteria? Discuss your findings with another student to see how far you agree.

b) From your own research, as well as Figure 11, draw an accurate annotated map of Gatwick's location, showing its situation and general accessibility.

c) Write a paragraph stating the ways in which you think Gatwick Airport's location is unsatisfactory. Try to include the perspective of travellers as well as other interest groups.

Figure 13   Gatwick Airport, 1936

The growth and development of the airport

There has been an airport at Gatwick since 1930, when a private aerodrome was opened (Figure 13). It was licensed as a public airport in 1934 under the name 'London South (Gatwick)'. During the Second World War, the airport was used by the Royal Air Force. In 1956, the old airport ceased to operate and construction of a new airport began on the present site.

The new airport re-opened in 1958. It was built partly on the site of the old Gatwick racecourse. Since then the airport has grown continually, and its facilities have had to be expanded. During the 1970s, the runway was lengthened to accommodate newer types of planes including wide-bodied aircraft like the Boeing 747 (jumbo jet), as well as Concorde. A circular 'satellite' departure and arrival area was opened in 1983 to replace the north pier, and there were improvements in terminal facilities.

In 1991, Gatwick became the sixth busiest international airport in the world (Figure 14) handling over 18 million passengers, some 202 000 tonnes of cargo and more than 163 000 aircraft movements. In 1987 Gatwick was briefly the world's second largest airport.

Figure 14   Gatwick airport, 1992

The airlines based at Gatwick have regular scheduled flights to 125 destinations (see Figure 15). Mark these locations on a world map.

Aberdeen
Abidjan
Abu Dhabi
Accra
Amsterdam
Antigua
Antwerp
Athens
Atlanta
Auckland

Bahrain
Baltimore/Washington
Bangkok
Banjul
Barbados
Barcelona
Beijing-Peking
Berlin
Bermuda
Berne
Billund
Bordeaux
Boston
Bremen
Brest
Brussels

Caen
Calgary
Casablanca
Charlotte
Cincinnati
Cleveland
Copenhagen
Dallas Ft-Worth

Deauville
Delhi
Denver
Detroit
Dortmund
Dubai
Dublin

Edinburgh
Edmonton
Eindhoven

Faro
Florence
Frankfurt
Funchal

Geneva
Genoa
Gibraltar
Glasgow
Grenada
Guernsey

Harare
Hong Kong
Honolulu
Houston

Ibiza
Innsbruck
Islamabad

Jakarta
Jerez de la Frontera
Jersey

Kano
Karachi
Kathmandu
Kingston

Lagos
Larnaca
Le Havre
Ljubljana
Los Angeles
Lourdes
Lyon

Maastricht
Madrid
Mahe Island
Malaga
Malta
Manchester
Manila
Mauritius
Miami
Milan
Minneapolis – St Paul

Montego Bay
Montpellier
Munster

Nantes
Naples
Newcastle
New York – JFK
New York – Newark
Nice

Orlando
Oslo

Paderborn
Paphos
Paris – CDG
Perpignan
Philadelphia
Pittsburgh
Pontoise
Port of Spain

Quimper

Rennes
Rome
Rotterdam
Rouen

Sanaa
San Juan
Seville
Sharjah
St Louis
St Lucia
Stockholm
Strasbourg

Toronto
Toulouse
Tunis

Valencia
Vancouver
Verona
Vienna
Vilnius

Zurich

Figure 15 Scheduled destinations from Gatwick airport

The airport's annual passenger capacity was increased to 25 million when a second terminal (see Figure 16) was opened in March 1988. The North Terminal was built to accommodate the huge increase in activity which had occurred at Gatwick. Since 1978, when the need for a second terminal was realised and planning began, the number of passengers using Gatwick each year has trebled from around just 7 million.

How does Gatwick Airport's role compare with that of other airports?

Gatwick's growth has not occurred without considerable planning. In 1983, the British Airports Authority (BAA plc) which owns the company now operating the airport, Gatwick Airport Limited (GAL), published a major planning document, called *Into the 1990s: The Gatwick Airport Master Plan Report*. In it, the strategy of developing Gatwick as a single runway, two terminal airport was affirmed. This would provide a passenger handling capacity of 25 million people (as we have already seen), and would allow those airlines already based at the airport to expand their services.

The role of Gatwick has always been as London's 'second' airport, so it has a smaller capacity than Heathrow with its four terminals and two runways. However, there are many other basic differences between Gatwick and other airports. These can perhaps best be considered under three headings: types of flight (scheduled or charter); passenger origin and destination; and models of aircraft using the airport.

A closer examination of the passenger figures given in Figure 10 shows that a substantial proportion of Gatwick's passengers in 1991/92 travelled on non-scheduled (or charter) flights when compared with those using Heathrow. The same is true now at Stansted, though the situation may change.

a) Calculate the percentage of charter (non-scheduled) passengers for Heathrow, Gatwick and Stansted in 1991/92 using Figure 10, and the percentage of scheduled passengers. Comment on your answers.

b) How might the different emphasis at Gatwick affect the airport's operation?

Figure 16 Gatwick's new North terminal

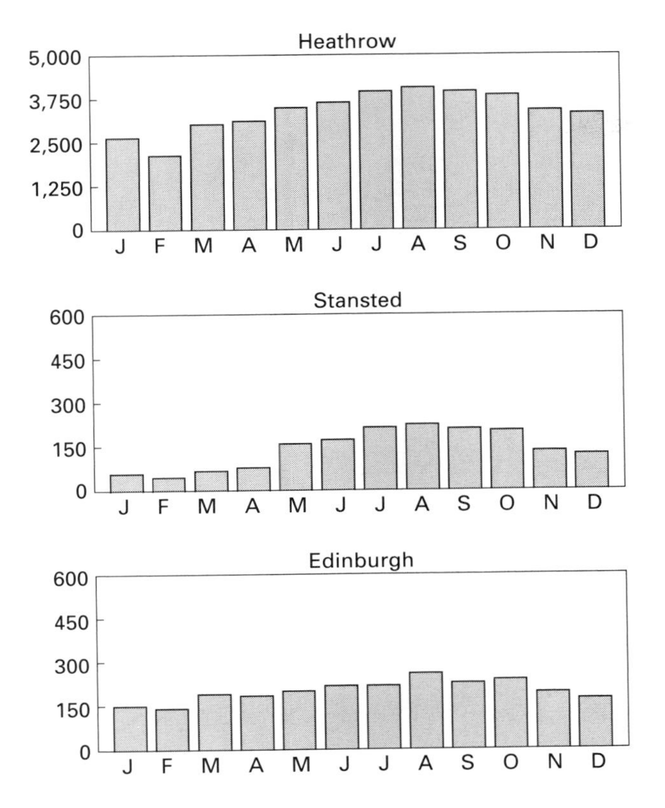

Aberdeen fixed wing

600
450
300
150
0
J F M A M J J A S O N D

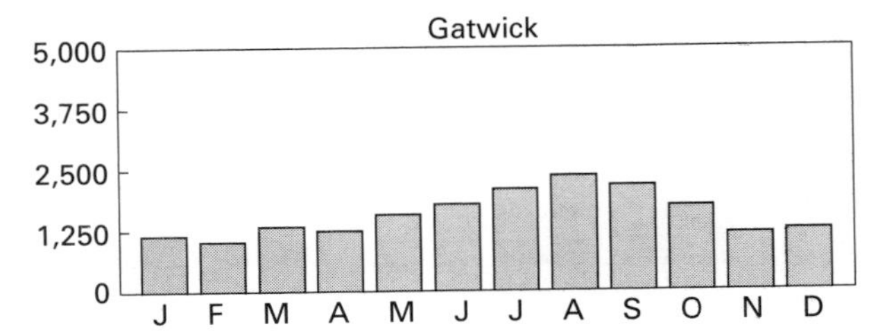

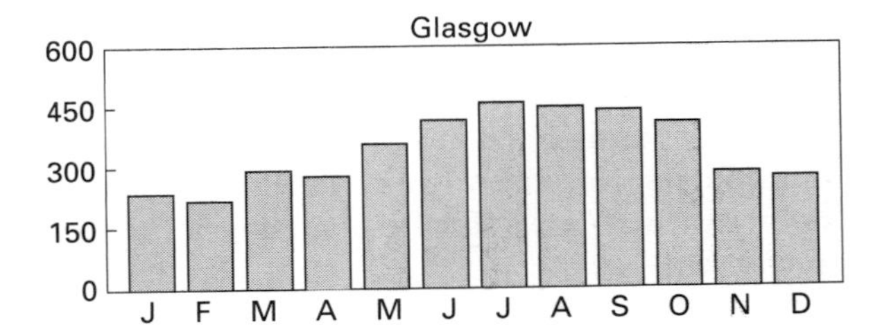

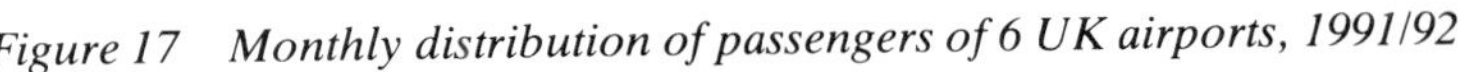

Figure 17 Monthly distribution of passengers of 6 UK airports, 1991/92

Source: BAA Traffic Statistics 1991

These figures reflect an important difference in the usage of the airports. Gatwick and Stansted do have scheduled services, but their principal role in the past was mainly to accommodate charter airline operations which service the multi-million-pound package holiday business. Until July 1991 Heathrow's traffic was almost entirely scheduled. Since that date non-scheduled services have been allowed to operate at Heathrow.

a) What evidence is there in Figure 17 that holiday traffic dominates air movements at Gatwick and Stansted?

b) Calculate the Passenger Flow Indices for Heathrow, Gatwick and Glasgow Airports:

$$\text{Passenger Flow Index} = \frac{\text{number of passengers in busiest month}}{\text{number of passengers in quietest month}}$$

c) What conclusions do you draw from your calculations and what implications do these indices have for the airports concerned?

d) What does a Passenger Flow Index equivalent to one mean?

The emphasis on charter operations at Gatwick is also reflected in the locations most frequently visited by aircraft departing from Gatwick. The leading destinations in terms of the number of passengers visiting them are Tenerife (Canary Islands), Malaga (Spain), Faro (Portugal) Los Angeles (USA) and Orlando (USA) and Paris, (France). Over half a million passengers travel to each of these destinations. All these flights are operated by airlines specialising in charter operations such as Britannia, British Caledonian and Monarch.

The situation at Heathrow Airport is rather different. Charles de Gaulle Airport in Paris is its most important destination, visited by over 2.6 million people in 1991/92. As a result of the well-developed scheduled 'shuttle' operations, there were over one million passengers carried to UK destinations like Glasgow, Edinburgh and Belfast in the same year. Manchester attracted almost as many passengers. In Europe the emphasis is on capital cities and business centres like Amsterdam, Dublin, Zurich, Frankfurt, Brussels and Geneva, rather than on Mediterranean resorts. The destinations outside Europe with over half a million visitors include New York, Los Angeles and Toronto. Longer-haul destinations like Tokyo also attract over half a million visitors.

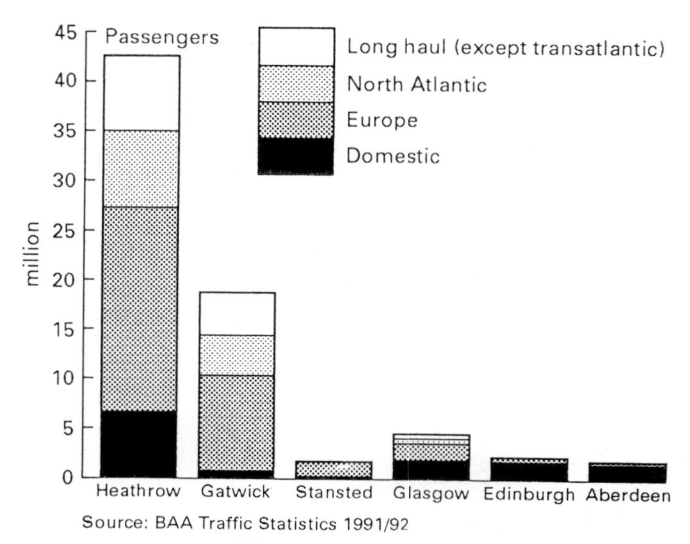

Figure 18 Origin and destination of passengers, 1991/92

a) Use Figure 18 to calculate the origin and destination of passengers using London's two principal airports as percentages. Now draw two divided bars to show this information. Write a short explanation of the differences you observe.

b) Using all the relevant resources you have seen so far, as well as Figure 19, compare and contrast the principal passenger flows at London's three airports. Consider the type and range of destinations served by each airport, and the relative numbers of passengers attracted to each of the principal destinations.

Another factor which may distinguish one airport from another is the sort of aircraft which use the airport. This is, of course, related to airline economics, which in turn are determined by the type and volume of passengers carried and the principal destinations visited. There are no simple rules about the sort of aircraft which suit each route, because the economics of different routes and the different bases of operation (scheduled flights versus charters) are complex. However, a basic distinction can be drawn between wide-bodied and long-haul jets on the one hand and narrow-bodied short-haul jets on the other.

Each airport's profile of aircraft movements by type is quite distinctive, as Figure 20 shows for the six airports operated by subsidiaries of BAA plc. The largest proportion of wide-bodied jets is found at Heathrow, which is the UK's flagship airport with the largest number of long-distance scheduled services.

	Terminal Passengers (000's)	% of total
Heathrow		
Paris (CDG)	2,679	6.4
Dublin	1,576	3.7
Amsterdam	1,523	3.6
New York (JFK)	1,507	3.6
Glasgow	1,292	3.1
Edinburgh	1,292	3.1
Belfast	1,175	2.8
Frankfurt	939	2.2
Manchester	904	2.2
Tokyo (NRT)	825	2.0
Brussels	804	1.9
Zurich	800	1.9
Los Angeles	709	1.7
Geneva	681	1.6
Rome	665	1.6
Milan	662	1.6
Munich	567	1.3
Dusseldorf	544	1.3
Toronto	540	1.3
Copenhagen	510	1.2
Other Routes	21,851	52.0
TOTAL	**42,045**	**100.0**
Gatwick		
Tenerife	665	3.5
Malaga	659	3.5
Faro	616	3.3
Los Angeles	609	3.2
Orlando	550	2.9
Paris (CDG)	549	2.9
Alicante	479	2.5
Palma	463	2.5
New York (EWR)	417	2.2
Atlanta	351	1.9
Miami	342	1.8
Amsterdam	329	1.7
Geneva	323	1.7
Malta	318	1.7
Boston	308	1.6
Houston	304	1.6
Hong Kong	299	1.6
Lanzarote	289	1.5
Frankfurt	278	1.5
Corfu	250	1.3
Other Routes	10,476	55.5
TOTAL	**18,874**	**100.0**
Stansted		
Dublin	235	12.3
Paris (CDG)	156	8.1
Amsterdam	112	5.8
Edinburgh	91	4.7
Palma	83	4.3
Glasgow	77	4.0
Channel Islands	75	3.9
Brussels	60	3.1
Stockholm	59	3.1
Aberdeen	57	3.0
Other Routes	913	47.6
Total	**1,918**	**100.0**

Source: BAA plc

Figure 19 Major passenger flow at London's principal airports, 1991/1992

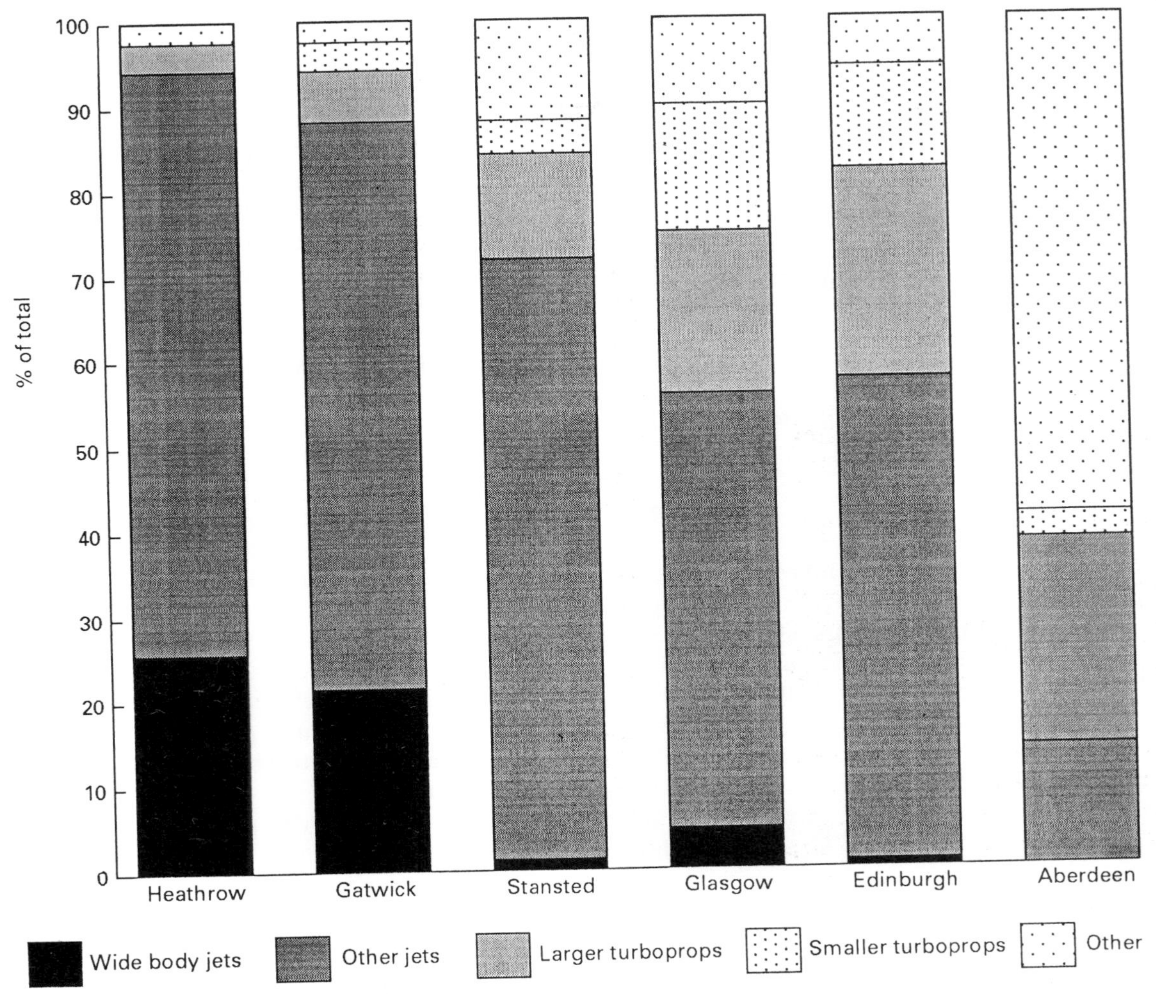

Figure 20 Air transport movements by type, 1991/92

Heathrow and Gatwick airports are used by different types of aircraft. The leading models at each airport are as follows:

Heathrow	%	Gatwick	%
Boeing 737	23.1	BAe 1–11	27.6
Boeing 757	14.5	Boeing 737	21.7
Douglas DC9	13.3	Boeing 747	5.9
Boeing 747	11.5	Douglas DC10	5.5

Narrow bodied jets like the Boeing 737, which mainly fly on short- and medium-haul routes, dominate air traffic at both airports.

Analyse Figure 20 carefully, by calculating the relative and absolute importance at different categories of aircraft at Heathrow and Gatwick. Try to relate your findings to the different roles of the two airports.

One final important role of an airport, which may distinguish it from others, is its function in relation to cargo. BAA policy is to increase cargo at all airports. The amount of cargo handled at Gatwick has increased from 50 000 tonnes in 1972 to more than 198 000 tonnes in 1991/92. This massive increase (charted in Figure 10) has meant that additional resources have been allocated to provide extra facilities for cargo handling at the airport. In 1977 a new Cargo Area was opened, comprising a transit shed, taxiway and apron. Fifteen planes can be accommodated on the Cargo Area apron, including the all-cargo version of the Boeing 747. More recently a meat inspection unit and cold store were added.

What is Gatwick Airport really like?

Most people's only point of contact with an airport is when they are leaving for or returning from a holiday or waiting to meet friends or relatives. In either case, we are unlikely to be able to make a rational evaluation of our surroundings. It is a useful exercise to visit an airport at another time, when a more objective assessment can be made.

Airport fieldwork

A visit to an airport to undertake fieldwork will provide you with a chance to explore the airport and learn about its organisation; to study incoming and outgoing flight patterns; to measure passenger flows; to evaluate airport amenities, and to look at the quality of the airport's environment. You will also be able to see the broader regional context within which the airport is located, e.g. the physical landscape, the nearby villages and other communities. You may also have an opportunity to talk to some of the people whose daily lives are directly affected by the airport.

Such a visit will have to be planned carefully just before it takes place. Here is a list of some of the activities you could include:

Orientation

Walk around the main concourse and make a topological sketch-plan of its layout. Mark clearly the arrival and departure areas, check-in desks, important passenger amenities, access points to public transport and car parks etc. A rough version only is required. You can redraw it later.

Arrivals and departures

At prearranged times go to either the departure or arrivals area. Examine the flight indicator boards. Note the place of origin and arrival time (or destination and departure time) of all flights in the set time period allocated. Compare your results with those of other groups. Does the pattern of destinations and places of origin vary with the time of day (or night)?

Passenger amenity survey

During your visit you should collect sufficient information about the number, type, location, range and quality of passenger amenities to be able to write a 600-word summary of your findings. Your report should be largely factual. It may be accompanied by maps and photographs.

Passenger count

At a prearranged time go to your allocated counting point:
a) junction of main concourse with the walkway from the car-parks
b) junction of main concourse with entrance from British Rail/Underground station
c) flight departure/passport control point
d) just outside the arrival area.
Count the number of people moving past your point in the predetermined direction in the time allocated (say 15 minutes). Compare your results with those of other groups, and plot all your results on one map or plan.

Feelings

Assess your feelings towards the surroundings in which you find yourself by completing Sheets 1 & 2 (Figure 21).

Environmental quality survey

Undertake an environmental quality survey of the airport's interior by completing Sheet 3 (Figure 21).

Try to visit a local airport if you are not able to visit Gatwick. It is important to experience the atmosphere for yourself, and to see its internal and external environment.

Having gone through the activities in this section, you should by now have a good idea what airports are like. You should also know about Gatwick's features, including those which distinguish it from other airports.

Evaluate Gatwick's current role as an airport by using all the information and resources available so far. You may use additional sources if you wish. Your task is to produce a 500-word article about the location and role of Gatwick. Imagine that this will be included in the new edition of the *Directory of World Airports*, which will be published soon.

Try to reach a conclusion about the importance of passengers and cargo. One map and one diagram may be submitted with your article.

Having looked at past trends, what do you think will happen to the volume of passenger and cargo traffic at Gatwick over the next decade or two? Will the airport continue to grow? How will other airport developments in London and the south-east affect its future? What impact will these trends have on the airport superstructure, its environment and on local communities? The next exercise investigates the statistics, to see whether expansion at Gatwick is likely to be necessary.

Sheet 1. Reaction sheet

Write down or draw something in as many of the following spaces as you can. Something you see in or around the airport which:

Pleases you	Frightens you	Relaxes you

Saddens you	Puzzles you	Amuses you

Sheet 2. Feelings about the airport's surroundings

These pairs of opposite words can be used to indicate your feelings about the airport environment. On each line, tick the space that best describes the places you are looking at e.g. is it: (a) very pretty; (b) pretty; (c) neither pretty nor ugly; (d) ugly; (e) very ugly?

	(a)	(b)	(c)	(d)	(e)	
bright	___	___	___	___	___	drab
clean	___	___	___	___	___	dirty
quiet	___	___	___	___	___	noisy
light	___	___	___	___	___	dark
pretty	___	___	___	___	___	ugly
smooth	___	___	___	___	___	rough
fresh	___	___	___	___	___	smelly
ordered	___	___	___	___	___	chaotic
interesting	___	___	___	___	___	boring
airy	___	___	___	___	___	musty
friendly	___	___	___	___	___	unfriendly
spacious	___	___	___	___	___	closed in
safe	___	___	___	___	___	unsafe
contrasting	___	___	___	___	___	uniform

Name: ______________________

Date: ________ Time: ________ Location: ____________

Sheet 3. Environmental quality survey

Site: ______________ Location: ____________ Date: ________ Time: ________

		+3	+2	+1	0	−1	−2	−3	
LITTER	No litter to be seen at all								Litter and rubbish everywhere
VANDALISM	No signs of vandalism								Buildings/plants seriously damaged
GRAFFITI	Buildings/walls free of graffiti								Graffiti covering most public surfaces
OPEN SPACE	Large areas of open space/ pleasant vistas								Atmosphere of overcrowding
LIGHTING	Lighting plentiful								Poor lighting. Dark corners.
FURNITURE	Sufficient benches/ litter bins etc. in good condition, attractive								Badly maintained benches/bins - unattractive and obtrusive
PLANTS	Plants and shrubs plentiful								Few plants or shrubs
SMELL	Air fresh, pleasant to breath								Atmosphere polluted or unpleasant
NOISE	Quiet atmosphere with pleasant, relaxing sounds								Loud, shrill sounds. Very disturbing.

SCORE: [ ] + points [ ] − points NET SCORE = [ ]

Figure 21 Fieldwork at airports

EXERCISE 4

Is there a good case for expansion at Gatwick Airport?

Over the last 50 years, as we have seen, the rise in the volume of air traffic using British airports has continued to increase. The two airports in the London area, namely Heathrow and Gatwick, seem to have attracted a disproportionate share of this increase. This is because the large airlines tend to be based in one or other of these two 'gateway' airports, which in turn means that they serve the greatest range of destinations with the greatest variety of flights. Both are accessible, with good transport links, though long journeys are still necessary for passengers living in the Midlands, the south-west and northern parts of Britain.

Forecasting the volume of future air traffic

All airports must respond to changes in the volume of air traffic, therefore it is essential that accurate forecasts of likely future trends are made. This is all the more essential in an area like south-east England, where the skies are already very crowded, and existing airport capacity is stretched to breaking point at certain times of the year. Popular concern about the quality of the environment makes accurate forecasts even more desirable.

Forecasting techniques need to be precise for two main reasons:

1 the construction of additional airport capacity, e.g. terminals, access roads, car-parking, runways etc., whether for passengers, airplanes or freight, is expensive. The use of such facilities has to be maximised if their construction is to be economic.

2 by its nature, the design and realisation of new terminal and runway capacity involves long 'lead-times'. Long-term forecasting is therefore required in order to make viable the huge investment required. This must be accurate for a period of maybe 10–20 years.

Since the late 1950s a series of forecasts have been made for air traffic growth, both for all the UK airports and for airports in specific regions. As air travel has risen in popularity these forecasts have become more frequent, and the techniques used more sophisticated.

Some of the forecasts made between 1956 and 1979 for air traffic growth in the London Area Airports can be seen in Figure 22.

Figure 22 Forecasts of traffic at London area airports, 1956–1990

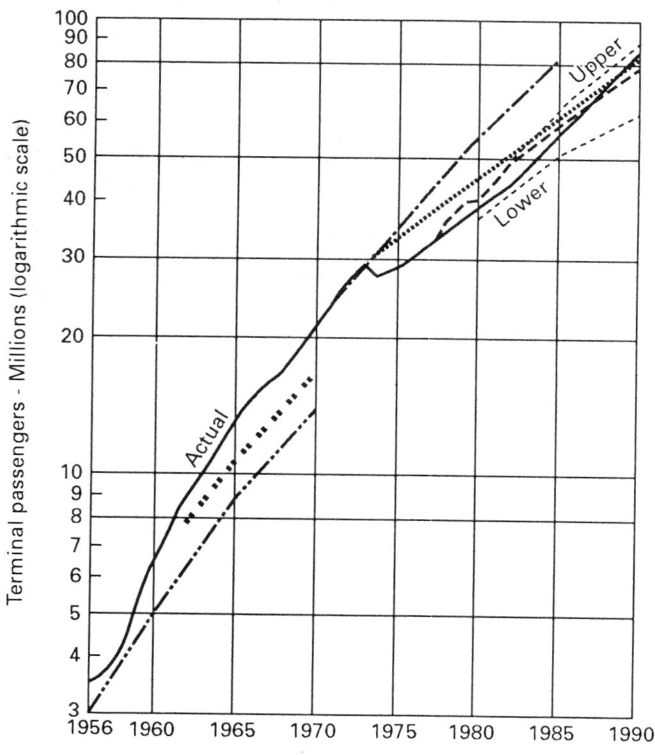

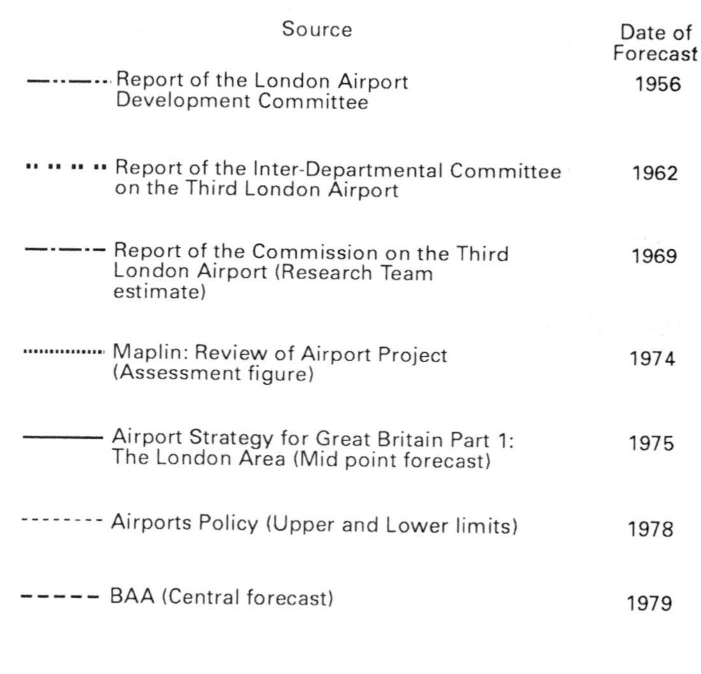

a) How did the estimates of air traffic growth between 1956 and 1970 compare with the actual growth recorded?
b) What factors do you think might have been responsible for any divergence?
c) Estimate the average predicted number of terminal passengers for 1989. Compare this figure with the actual number of passengers using Heathrow, Gatwick and Stansted in 1988/89 (from Figure 10). Has the accuracy of forecasting techniques improved?

During the 1980s estimates of air traffic growth continued to be revised upwards. In December 1988 the Department of Transport published forecasts of passenger air traffic at UK airports up to the year 2005 (Figure 23). These suggested strong growth in demand both at London and regional airports. The Secretary of State for Transport noted that the figures predicted higher traffic volumes than those estimated in 1986.

Until then it had been generally believed that although Heathrow and Gatwick were operating at near-capacity for much of the time, their passenger throughput would continue to increase. This would be the result of a continued growth in the size of aircraft and an increase in off-peak uses of the airports (called peak-spreading). Furthermore, there was known to be substantial spare runway capacity at Stansted, where terminal facilities under construction would lead to an increase in passengers from 1 million to 8 million initially, and then perhaps to 15 million each year, by the end of this century.

The 1988 forecast (see Figure 23) led to a re-evaluation of the government's 1985 White Paper entitled *Airports Policy*. Here the government concluded, in respect of London airports, that developments then in hand or planned would give enough capacity to meet demand into the mid-1990s. The Civil Aviation Authority were asked to make formal recommendations about the UK airport capacity needed to cope with demand through to 2005.

What factors do you think need to be taken into account when trying to predict air transport growth? List the important factors, then compare your answers with others in the group.

In order to predict future air traffic growth several assumptions have to be made. Among the critical factors are the assumptions about future economic growth (in the UK, Europe and the rest of the world). Economic growth directly affects people's affluence and personal disposable income, which in turn influences the choices they make about travel and holidays. Assumptions also have to be made about the relative growth and attractiveness, in terms of price, speed and convenience of competing forms of transport.

There are many uncertainties, such as the effect on air travel of the Channel Tunnel, and the further development of a high-speed rail network across Europe. The future trend in the price of air fares is also unclear. They may rise to reflect the increasing price of oil or the damage done to the environment, or they may fall in response to further liberalisation. And how will the tourist industry develop in the next 15 years – is the age of package holidays over?

Despite all these difficulties in forecasting, a number of groups make their own estimates of what is likely to happen. Officials at the Department of Transport carry out this vital task for the government, who also take advice from the Civil Aviation Authority (CAA). The airport owners and operators also make their own estimates. The largest of these is the British Airports Authority (BAA plc), which operates six airports (including Heathrow, Gatwick and Stansted) through its

	1987	1995		2000		2005	
	Base year	Low	High	Low	High	Low	High
London area airports**	57.4	77.9	87.9	88.9	113.6	101.4	147.1
Regional airports	28.6	42.1	51.0	52.4	68.2	62.6	87.4
UK total	**86.0**	**120**	**138.9**	**141.3**	**181.8**	**164.0**	**234.5**

**Heathrow, Gatwick, Luton, Stansted, London City

Figure 23 Department of Transport forecasts of future UK air traffic growth (millions of passenger movements), 1988

	1990	1993		1995		2000		2005	
	Actual	Low	High	Low	High	Low	High	Low	High
PASSENGERS (millions)									
London Area*	64.8	67.0	74.1	70.0	80.8	85.6	105.6	100.5	131.4
Southampton	0.5	0.5	0.6	0.5	0.8	0.7	1.0	0.8	1.4
Scotland**	8.7	9.1	10.2	9.6	11.2	11.5	14.2	13.3	17.5
Total	74.0	76.6	84.9	80.1	92.8	97.8	120.8	114.6	150.3
AIR TRANSPORT MOVEMENTS (000's)									
London Area	579	607	625	625	645	690	740	775	833
Southampton	20	19	25	22	33	25	38	28	43
Scotland	202	206	215	208	227	220	242	227	264
Total	**801**	**832**	**865**	**855**	**905**	**935**	**1,020**	**1,030**	**1,140**
CARGO (000's of tonnes)									
London Area	951	1,040	1,100	1,080	1,200	1,180	1,500	1,350	1,910
Southampton	1	1	1	1	2	1	2	1	2
Scotland	26	28	33	29	35	32	42	34	50
Total	**978**	**1,069**	**1,134**	**1,110**	**1,237**	**1,213**	**1,544**	**1,385**	**1,962**

* Heathrow, Gatwick, Stansted
** Glasgow, Edinburgh, Aberdeen

Figure 24 Long-term traffic forecasts for BAA airports

subsidiary companies. Their latest predictions (see Figure 24) show that traffic at BAA airports will double by the year 2005. They claim that passenger numbers will rise from 74 million (1990) to between 114.6–150.3 million (2005), and that cargo tonnage will double.

When all these forecasts are viewed together (see Figure 25) similar trends emerge. Predictions about increased air traffic are only meaningful, however, when compared with airport capacity. Current airport terminal capacity at London area airports (existing or planned) is around 76 million passengers per year (38 million at Heathrow, 25 million at Gatwick, 8 million at Stansted and 5 million at Luton). By the end of the century, the four principal London area airports could take 100 million passengers a year (Heathrow: 50m; Gatwick: 30m; Stansted: 15m; Luton: 5m.) It seems likely that there will be a serious shortfall after this.

a) In what year will airport capacity in the London area be exceeded by the number of passengers forecast?
b) By the year 2000 how much extra capacity will be required?

Estimating how serious the problem will be is a complicated issue, as there is no single appropriate measure of capacity. Statistics relating to runway capacity, terminal capacity and airspace capacity must all be considered. However, it is clear that with the BAA, CAA and the Department of Transport all predicting average annual growth of over 4%, the numbers of passengers forecast to use London area airports will probably exceed airport capacity before the year 2000, and more runway capacity will be needed after that date.

Figure 25 Long-term passenger forecasts at London area airports

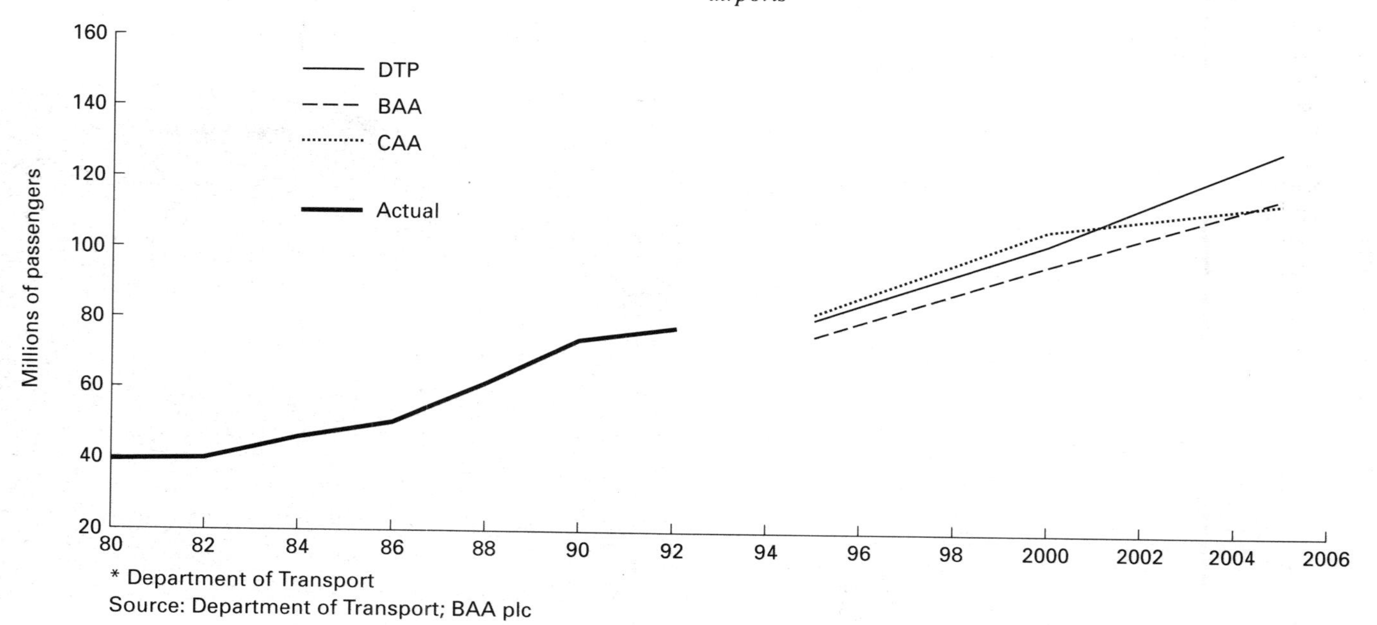

Demand continues to grow at Gatwick

The figures we have examined so far appear to suggest that air traffic will continue to grow in the London area, requiring more terminal and runway capacity to be constructed soon.

When the situation at Gatwick is examined independently the figures confirm that it has been one of the fastest growing airports in the United Kingdom. An average annual increase of 10% in the number of passengers handled between 1980 and 1990 (Figure 19) has made Gatwick the world's sixth busiest international airport in terms of the volume of passengers. This is despite a decline in passenger numbers in the early 1990s, due to the combined effects of the recession, worries over terrorism during the Gulf War, and the opening of new terminal facilities at Stansted.

Traffic growth at Gatwick (see Figure 26) can be attributed to a number of factors:

1 continued expansion of the market for holidays abroad, particularly in the Mediterranean, and the consequent demand growth in charter airline operations;

2 the development of low-cost transatlantic air traffic based at Gatwick, first with Laker Airways, then People Express (both no longer operating) and now with Virgin Atlantic (Figure 27);

3 the continued development of Gatwick's scheduled services, especially to Europe, the Middle East and the Far East.

The airport's excellent location and consequent good accessibility are certainly additional factors which have contributed to its popularity and its growth. Gatwick, unlike Heathrow, is situated in the Sussex countryside well outside the built-up sprawl of London. However, it is easy to reach by road and rail, and in many ways is less congested than the larger airport at Heathrow.

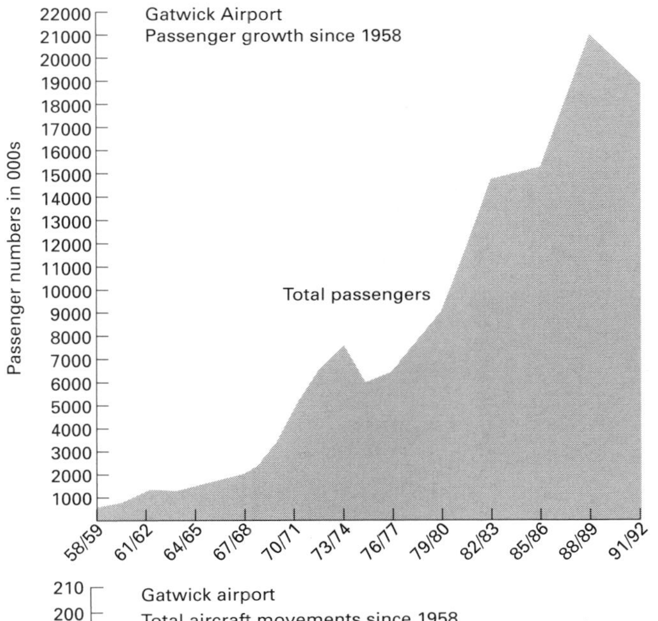

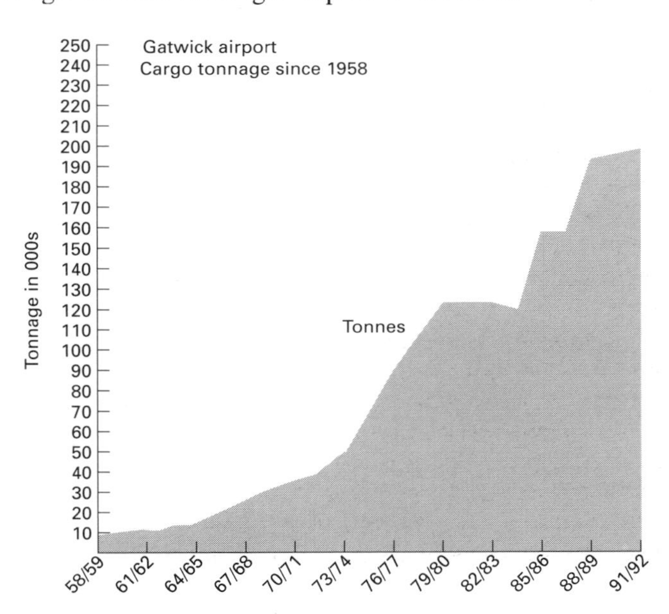

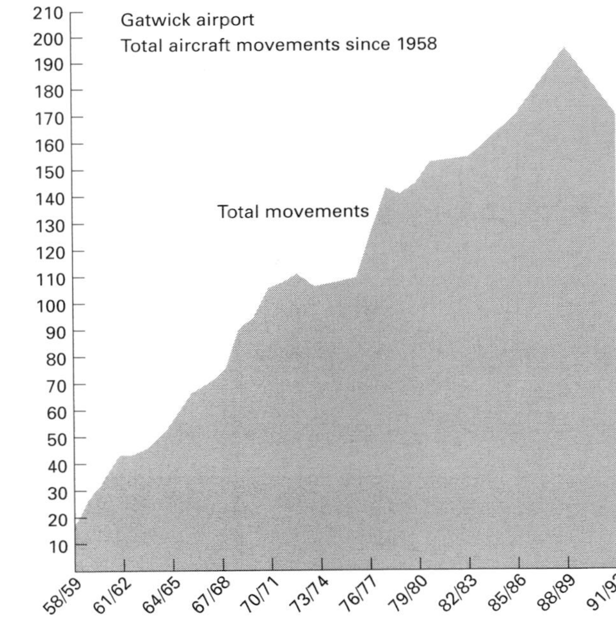

Figure 26 Changes at Gatwick since 1958

Figure 27 Virgin Atlantic – a new carrier

a) In small groups discuss the possible solutions to the likely future problems suggested by the air traffic forecasts presented in Figures 23 to 25.
b) Try to rank the various alternatives according to their desirability.
c) One suggestion (see Figure 28) is that facilities at RAF Manston in Kent should be expanded. Make a list of points you could use to argue:
(i) in favour of expansion at Manston
(ii) against expansion at Manston.
Use additional books and resources to help you research the arguments and compile your lists.

Forlorn battle base favourite to expand

MANSTON airfield in Kent has a slightly forgotten, forlorn air. The civilian side of this famous Battle of Britain base has seven old Boeing 707s laid up on the apron and a cluster of light aircraft from the local flying club.

This summer there will be eight charter flights a week to a few Mediterranean destinations and the Channel Islands. About 30,000 passengers will pass through Manston in 1989 — compared to 40 million at Heathrow.

The military side is also far from hectic; Manston's main role is as a Nato diversion airport. There is an RAF fire-fighting school, a flight of Sea King rescue helicopters and a few Chipmunk trainer aircraft. Yet Kent International Airport, as it has been renamed, could be London's fourth major airport by the end of the century.

A leading merchant bank has already shown an interest in investing in Manston's expansion following the CAA report. BAA and Manchester Airport's annual results show that large airports can be highly-profitable businesses. The bank has made contact with Seabourne Express, the parcels and courier company which has a 125-year lease on passenger operations at Manston and has embarked on a limited expansion programme.

Today, Kent International aspires to be London's fifth airport, behind Heathrow, Gatwick, Stansted and Luton. But it may overtake Luton around the turn of the century.

Manston's greatest asset is a 9,000ft runway, long enough to take the largest wide-body jetliners. There is room for a 1,000ft extension which would allow long-range Boeing 747 flights.

There are two other reasons why Manston is the front-runner for expansion outside of London's big four airports. First, the airfield is owned and run by the RAF. If the Government accepts a new London airport is needed, it can dictate Manston's future and make a great deal of money by earmarking it for private sector expansion.

Second, Manston fits in well with the requirements of air traffic controllers, who will have their say in deciding where the growth will go. It is outside the highly-congested Terminal Manoeuvring Area, the block of airspace over the Home Counties reserved for jets using the four London airports. It is also on the south-east edge of England, where most air traffic growth will be. Kent International sits on a bleak, flat area of land with sea on two sides and a power station on another. The open, thinly-wooded countryside in this part of Kent is not of the "Garden of England" variety.

It is 70 road miles from the capital, considerably further than any of the four London airports, but major road improvements linking Manston to the nearby M2 are under way.

A rail line passes within a mile, so a new station linking Manston to the planned high-speed Channel tunnel track would bring the airport within an hour of central London.

Although Kent is already burdened by enormous development pressures, rapid growth at Manston could face less environmental opposition than at any other airport in the South-east. It is surrounded by countryside, and for much of the time when the aircraft are noisiest — immediately after take-off and on approach to landing — they can be routed over the sea.

But Ramsgate, two miles east of the airport, would suffer. New houses are being built one mile from the runway's eastern end and jets coming in to land would usually sweep low over the town.

Thanet District, which includes the airport and Ramsgate, Broadstairs and Margate, has male unemployment of 15.4 per cent, more than twice the national average.

The area has been hit by the decline of the local coalfield and there are fears that further jobs will be lost when the Channel tunnel takes trade from the ferry ports of Ramsgate and Dover.

Thanet District Council and Kent County Council are firm supporters of expansion — up to a point. Ian Gill, Thanet's chief executive, said: "We're not prepared to completely sacrifice our quality of life here for economic growth."

Thanet is preparing its own document for the CAA explaining how large it wants the airport to grow. Mr Gill said it was too early to give any figures, but it would be more than a million passengers a year.

Tony Hart, Kent County Council's Conservative leader, said: "We don't want to go over the top and reach a horrendous situation like Gatwick.

"We want the best of both worlds — an international airport for Kent which would help our real unemployment blackspot without the overdevelopment which would harm the environment."

At Manston, a new £1m terminal will be finished this summer, capable of handling one million passengers a year. It is a gesture of faith by Seabourne Express and is financed partly by soft loans from the European Community made available because of the pit closures.

Even with the rosiest of scenarios, it will take at least four years before Kent International attracts that many passengers.

Whether Thanet and Kent will be able to ride the tiger of airport growth once Manston soars above that level remains to be seen. Heathrow and Gatwick have shown how, once a successful airport reaches a certain size, there are enormous pressures for it to keep on expanding.

The communities and councils around Britain's two largest airports are saying it is time to call a halt; the CAA and BAA are forecasting that they will grow much busier.

A Supermarine Spitfire, an emblem of RAF Manston's illustrious past.

Figure 28. Source: *The Independent* 27 February 1989

Crowded airports and the skies above them are a major cause for concern. This concern recently led to an investigation by the House of Commons Select Committee on Transport. Their report on Air Traffic Control Safety was published in 1989. In their deliberations they took evidence and advice from all the interested parties, and produced a report with some interesting observations and conclusions:

- '. . . there is little doubt that decisions, counter-decisions and vacillation by previous governments have left the London area with inadequate runway capacity.'
- '. . . the demand for air traffic movements is likely to exceed runway capacity by or before the turn of the century.'
- '. . . in aviation terms, it is most undesirable that an international airport should have only one main runway.' (This is the case at Gatwick and Stansted.)
- '. . . if a decision is taken to build a new runway to serve London the timescale is important. BAA . . . believe that a lead time of up to 10 years would be needed after a decision to make an application.'
- '. . . we recommend that, in spite of the many difficulties and obstacles and the inevitable controversy, all efforts should be made to provide a second main runway at Gatwick.'

This last suggestion has naturally aroused strong feelings, especially near the airport. One reason is that the construction of a new runway at Gatwick would break a legally binding agreement between the British Airports Authority and West Sussex County Council, entered into in 1979, which stated that there would be no second runway at Gatwick for a period of 40 years.

Would expansion at Gatwick benefit the area?

Gatwick Airport is the largest employer in Sussex, offering a range of jobs to skilled technicians and highly trained pilots at one end of the spectrum and to unskilled cleaning and catering staff at the other. More than one-third of the workforce is female, and there is considerable scope for young people to gain either full-time or part-time (seasonal) employment. A total of almost 22 000 people are employed directly at the airport.

The existence of the airport also provides employment opportunities beyond the airport itself. In other words, some jobs are directly related to the airport being there, but are not actually located on-site. Airport-related employment falls into four categories:

1 On-airport

Employment located at the airport, with the exception of hotels

2 Airport-associated

Employment whose existence is directly related to the operation of the airport, and which, while not located at the airport, requires of necessity a location close to it

3 Intermediate

Employment exclusively engaged in providing goods and services to activities directly related to the operation of the airport and not essentially located close to the airport

4 Secondary

Employment supported by the expenditure of employees in the on-airport, airport-associated, and intermediate categories, and their families.

All categories of employment have continued to grow as Gatwick has expanded. The neighbouring town of Crawley has one of the lowest unemployment rates in Great Britain as a result.

a) How significant is Gatwick Airport as a local employer in West Sussex?
b) What categories of employment are represented by the five job advertisements in Figure 29?
c) Make a list of different jobs in each of the employment categories listed above, and present it in tabular form.

Figure 29 Some jobs available at Gatwick

In total, the airport is estimated to have created employment for some 200 000 people both directly and indirectly in the Sussex/Surrey area.

The presence of the airport has benefited the Gatwick area through what economists know as the 'multiplier effect'. The net result is that Gatwick Airport has been a significant stimulus to the regional economy. Through the multiplier, income generated by the airport (wages, company profits and airport revenues) is recycled in local shops and businesses, helping the whole area to benefit economically from the airport's presence. This improves the social and economic infrastructure and makes the area more attractive for industrial and other forms of development (Figure 30).

In what other ways has the airport at Gatwick helped to shape the geography of south-east England?

Some might argue that to preserve Gatwick as 'the airport in the countryside' the British Airports Authority has actually improved the environment. Certainly the environment has been altered by landscape architects. BAA's objectives are to 'conserve wherever possible and to supplement existing landscaping resources, to minimise the visual impact of the airport on the immediate rural setting and to counterbalance extensive building developments'. Thousands of trees have been planted, and existing hedgerows and trees have often been incorporated into schemes designed to enhance the environment.

Do you agree with the suggestion that airport authorities do actually enhance the physical environment around airports through landscape architecture and tree-planting schemes? Try to research this proposition in relation to your nearest large airport, and explain your views fully.

There is as much debate about the harmful impact that airports have as there is about their benefits. These will be examined in the next section.

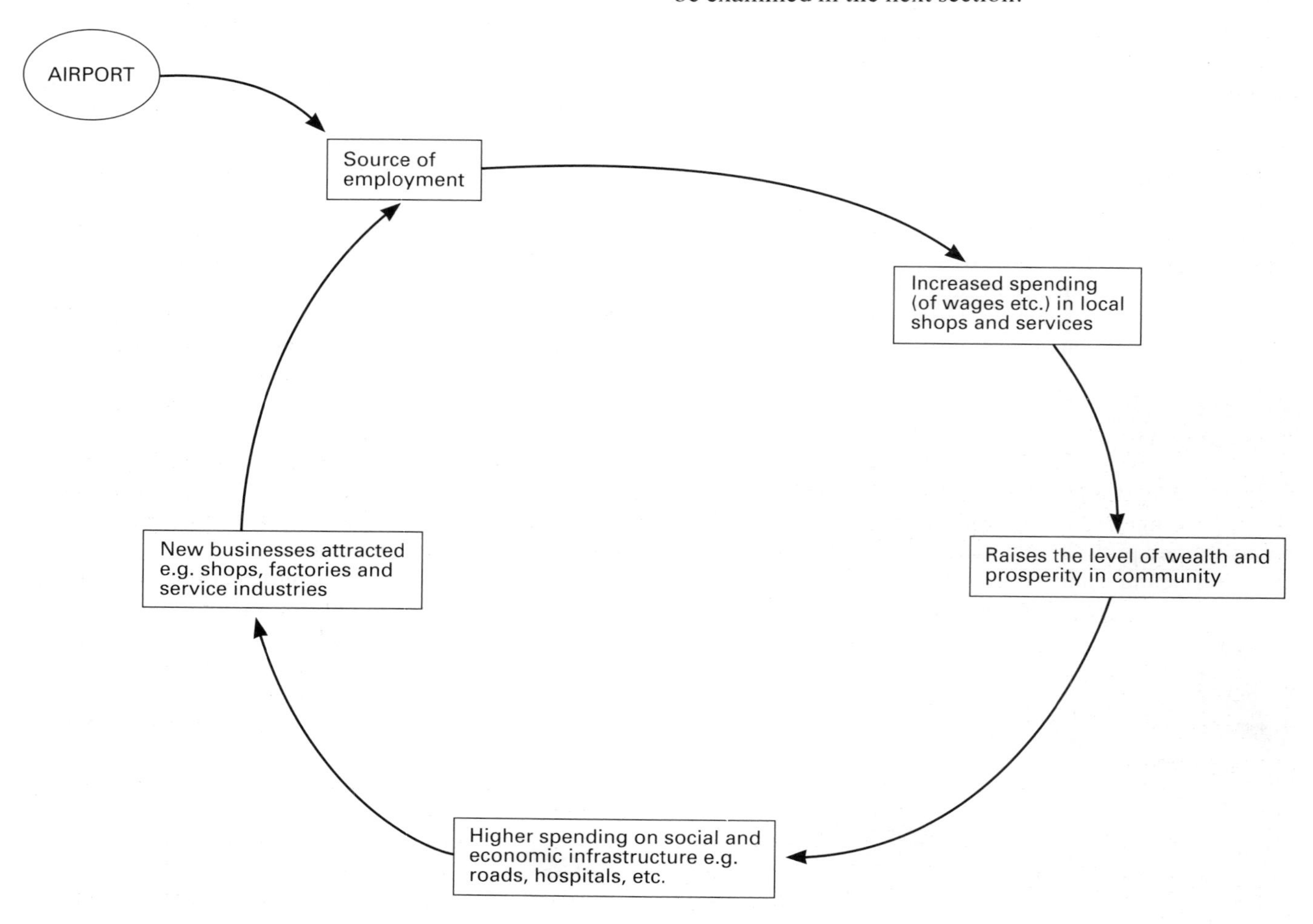

Figure 30 Gatwick's local 'multiplier' effect

EXERCISE 5

Does Gatwick Airport affect local people and environments?

The effects of Gatwick Airport on local people and environments was the main focus of a public inquiry convened in 1979. Its objective was to decide whether to construct a second passenger terminal at Gatwick. Among those people who gave evidence were local residents, environmentalists, scientists, local councils, amenity groups and the airport authorities. All these groups and many others have particular views about the likely effects of airport expansion. At the end of the inquiry, permission was given to the British Airports Authority to extend Gatwick's terminal facilities. A second terminal, called the North Terminal (see Figure 31) was opened in 1988.

The main objections to airport expansion at Gatwick fall into three main groups: those concerned with noise; those relating to pollution and the consequent damage to vegetation, and a final group comprising all the other effects. This exercise will allow you to examine some of the evidence in each of these groups.

Figure 31 The new North Terminal at Gatwick

Airport noise

One of the most obvious impacts experienced by people living near airports is the increase in environmental noise levels. Aircraft by definition create a certain amount of noise, especially on take-off and landing. At its best this noise constitutes a nuisance to local residents, but at worst it can ruin life for people by causing almost constant annoyance during daylight hours, particularly for those living under the main flight paths.

In Britain, the generally accepted way of measuring and quantifying airborne aircraft noise is by the 'Noise and Number Index' (NNI). This measurement originates from a study done at Heathrow Airport in 1961, and has been in general use since 1963. NNI is derived from a mathematical index, which takes account of the average noise level heard expressed in perceived noise decibels (PNdBs) and the number of aircraft movements. The Noise and Number Index effectively describes the average community reactions to air noise disturbance. Although there is some debate about the index, an NNI of 35 has become equated with low annoyance from noise, and an NNI of 55 with high annoyance.

The Noise and Number Index has been criticised for a variety of reasons:

1 It only considers the noise from aircraft in flight, and takes no account of ground noise, such as that generated by the ground running of engines, taxiing aircraft, 'start of roll', and reverse engine thrust.

2 NNI measurements do not include aircraft movements during the evening and at night, when aircraft noise poses more of a disturbance. The calculation of NNI only includes daytime summer aircraft movements.

3 People have become generally less tolerant of noise since 1961 when the NNI was evolved. Concern about noise and other environmental factors has grown.

4 The Noise and Number Index is unique to the United Kingdom. Other countries base their aircraft noise indices on the equivalent continuous sound level (Leq) scale. The UK Aircraft Noise Index Study (1985) suggested that a Leq-based index might be more useful than NNI for assessing the noise impact of airport development at rural sites.

5 The original social survey on which NNI is based was carried out near Heathrow Airport. The criteria could be different for other airports, such as Gatwick, in different types of location.

Draft a letter to the airport authorities voicing criticism of the method by which noise is measured. It should show your understanding of the criticisms levelled against the Noise and Number Index.

The spatial distribution of noise around Gatwick can be shown cartographically on maps using 'noise contours'. Figure 32 shows the noise levels around Gatwick Airport in summer 1988. Many communities are clearly affected by noise from the airport above the 'low annoyance' level of 35 NNI. However, the airport authorities state that Gatwick's single runway presents a natural limit to aircraft activity, and claim that as aircraft engines have become increasingly quieter, so the disturbance from airborne noise at Gatwick has declined over recent years.

a) Describe the pattern shown by the noise contours (Fig 32) and explain their east–west orientation.
b) Which communities are most severely affected by noise?
c) Comment on the likely impact (intrusion) of airborne aircraft noise on the sort of communities likely to be found in the West Sussex and Surrey areas around Gatwick.

There are, of course, restrictions on the amount of noise that aircraft can make during take-off and landing, and a system is in place to monitor noise levels. Gatwick has four monitoring points, or 'electronic ears', one each at Rusper, Capel, Smallfield and Copthorne (Figure 33). Each monitoring point comprises a microphone mounted on a pole about six metres above ground level and linked to a central recording unit. Every time an aircraft approaches, the microphone is automatically switched on. At each site, all noise above a certain level is recorded together with its exact time and duration. These recordings are then checked to see if any aircraft have broken the noise restrictions that apply to planes taking off from Gatwick. They stipulate that aircraft shall not cause more than 110 PNdB (perceived noise decibels) by day, and 102 PNdB at night at the relevant monitoring point. Aircraft must also have reached a height of 300 metres by the time they pass over these monitoring points.

The monitoring equipment lies close to the Standard Instrument Departure Routes (SIDs). These are also shown on Figure 33. Only outgoing aircraft are bound by noise regulations. Incoming aircraft are exempt from such restrictions so that pilots are free to use engine power as necessary to achieve a safe landing.

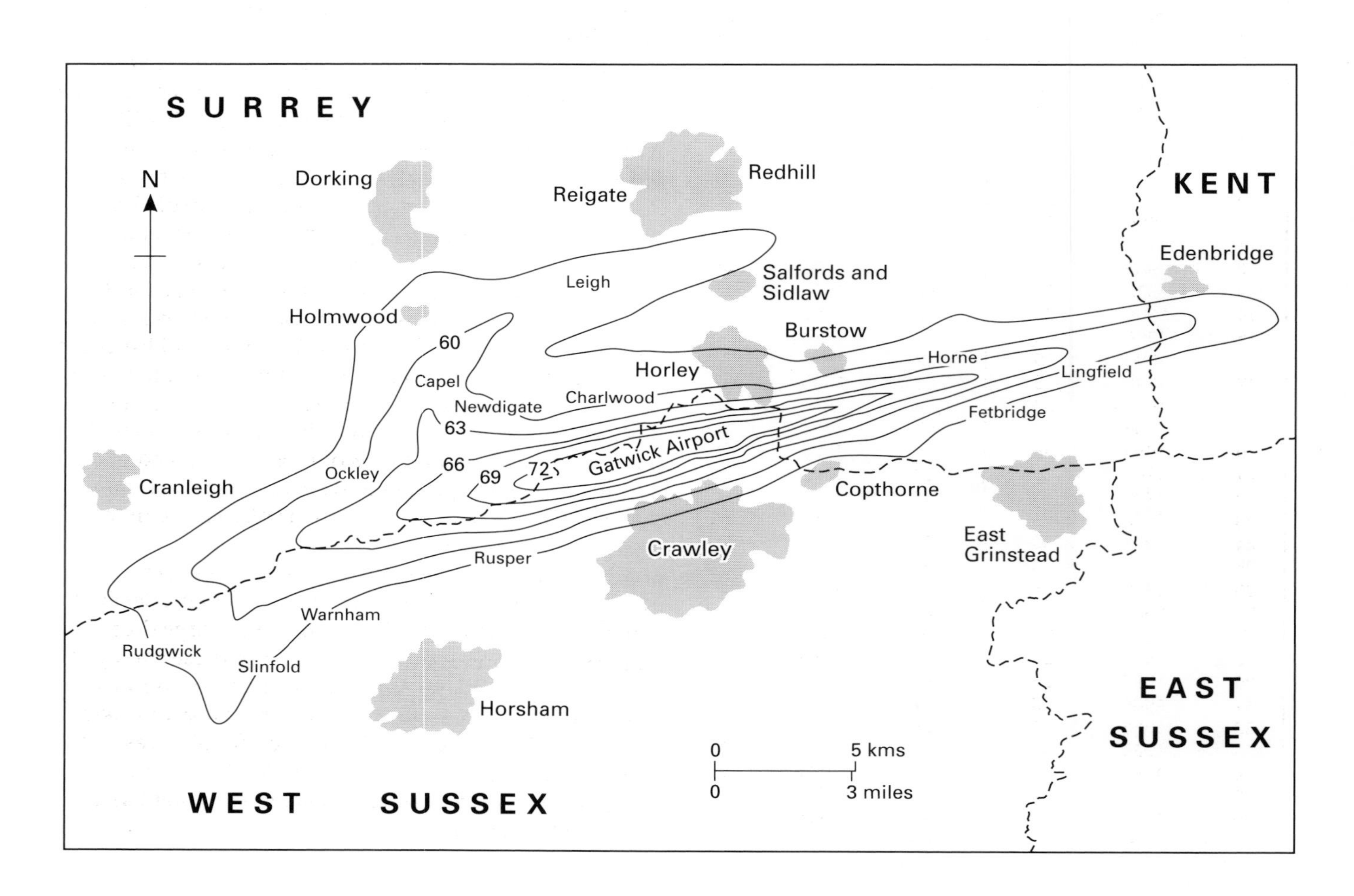

Figure 32 Noise contour map for Gatwick, Summer 1988

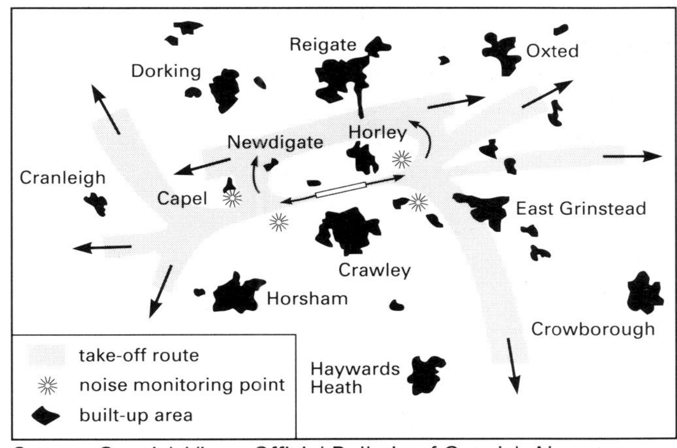

Source: Gatwick Views Official Bulletin of Gatwick Airport Consultative Committee circa 1981

Figure 33 Take-off routes and noise monitoring points

Data for isoline map of airport noise around Gatwick Airport

Map no.	Year	NNI	Map no.	Year	NNI
1	1985	35	37	1986	35
2	1985	35	38	1985	35
3	1985	35	39	1986	35
4	1986	35	40	1986	35
5	1985	35	41	1985	35
6	1986	35	42	1985	35
7	1986	35	43	1985	35
8	1985	35	44	1985	35
9	1985	35	45	1985	35
10	1986	35	46	1985	35
11	1986	35	47	1986	35
12	1985	35	48	1985	50
13	1985	35	49	1986	50
14	1986	35	50	1986	50
15	1986	35	51	1985	50
16	1986	35	52	1986	35
17	1985	50	53	1985	35
18	1985	50	54	1985	35
19	1986	35	55	1985	35
20	1985	35	56	1986	35
21	1985	35	57	1986	50
22	1986	35	58	1985	50
23	1985	50	59	1986	35
24	1985	50	60	1985	35
25	1986	35	61	1985	35
26	1985	35	62	1986	35
27	1986	35	63	1985	35
28	1986	35	64	1985	50
29	1986	50	65	1986	35
30	1986	35	66	1985	35
31	1986	35	67	1986	35
32	1986	35	68	1985	35
33	1985	50	69	1986	35
34	1986	50	70	1986	35
35	1986	50	71	1985	35
36	1985	50	72	1985	35

Figure 34 Data for isoline map of airport noise around Gatwick airport

Noise and Number Indices (NNI) for locations around Gatwick Airport are contained in Figure 34 for 1985 and 1986.

a) Using the data provided, make a tracing overlay of Figure 35. On it plot the noise contours for 35 and 50 NNI for 1985. Shade the area of maximum noise impact.

b) Repeat the exercise on a new overlay, this time using the data for 1986. Again shade the area of maximum noise impact.

c) Describe any differences in noise levels around Gatwick in these two years.

The airport authorities continue to claim that noise levels around Gatwick are declining. The Civil Aviation Authority estimated that in 1978 around 35 000 people lived inside the 35 NNI contour, and that by 1991, with a second terminal in operation, the number of people living within this contour would be halved to around 17 000.

Noise levels may indeed be declining, but the nuisance from noise remains a real one to those living close to the airport. The 'nuisance' caused by the airport is recognised by the existence of a system of grants. Those properties which are 'severely' affected by aircraft noise can be sold to the British Airports Authority if their owners wish. This applies only to properties inside the 65 NNI contour, and to those people whose ownership dates back to 1969.

Since 1980, grants have been available for sound insulation works to be carried out on properties inside an area based on the 50 NNI contour and the 95 PNdB footprint. This represents some 650 dwellings. The scheme is to be extended to cover parts of Charlwood and Tinsley Green as well as the southern tip of Horley. However, local lobby groups want this area extended further to all properties inside the 40 NNI contour.

Noise caused by aircraft at night is controversial. In 1986 the results of a study into *Noise Disturbance near Heathrow and Gatwick Airports* (1984) was published. It surveyed five sites to try to establish the extent to which sleep was being disturbed by aircraft noise. Two of the sites were near Gatwick, at South Horley and Lingfield. Here between 30% and 43% of those questioned reported difficulty getting to sleep due to aircraft noise and between 22% and 25% were awoken more than once each night by it.

The report indicated that noise levels between 11.30 p.m. and 6.30 a.m. improved by about 2 decibels between 1979 and 1984. There was also a large drop in reported air noise wakings from 40% of those surveyed at Lingfield in 1979 to 17% in 1984. Traffic noise was cited as frequently as aircraft noise as a cause of waking. It was concluded, however, that aircraft noise was still the main cause of disturbance at night. A 25% increase in 'quiet' aircraft movements at night was sought as a result of the report.

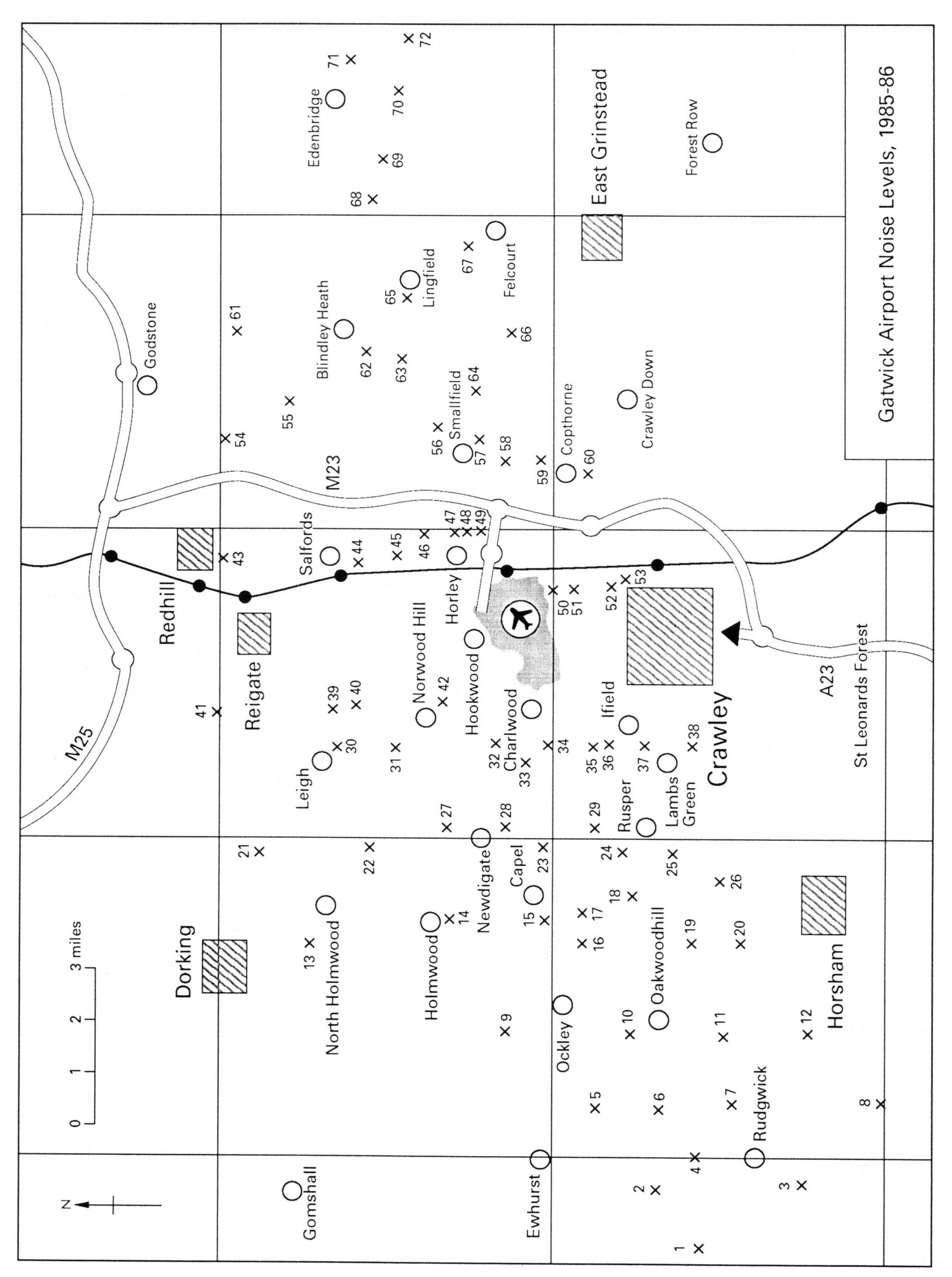

Figure 35 Gatwick airport noise levels, 1985–86 – Base map

a) Look at the documents in Figure 36. What was the general reaction to the Night Noise Study?
b) From the evidence you have seen, do you think there is a serious noise problem at Gatwick?

Any suggestions of increasing the number of night flights are strongly resisted by local residents, pressure groups and by the Gatwick Airport Consultative Committee. There is clearly noise from the airport, and it does disturb people. What can be done? There are perhaps four main options: to extend the insulation grants scheme so that the effects of noise on local residents are lessened; to quicken the introduction of quieter aircraft so that noise is reduced at source; to introduce an environmental levy forcing airlines to pay for the disturbance they are causing (the polluter pays principle), or to ban all night flights.

Analyse the advantages and disadvantages of each of these measures. Then work out an ideal and realistic scheme for noise reduction in airport environments. Try to take into account the viewpoints of all the interested parties.

Gatwick Area Conservation Campaign

GACC

NEWSLETTER NO.26 **JULY 1986**

NIGHT FLIGHTS - Another Government study

This document was issued in May. While the Study shows that disturbance has reduced, it also confirms that the situation remains extremely unsatisfactory.

We have argued strongly that Government policy should be one of continuing environmental improvement and that the policy advocated in GACC's statement "Good Neighbour Airports" should be adopted. This would involve a gradually increasing "core" night period when no night flights are allowed, together with a tapering in and out of morning and evening flights.

We have also stressed that there is no justification whatsoever for Gatwick having more night flights than Heathrow - indeed on the basis of the Study it should be the reverse, since the Study shows that people near Gatwick are more disturbed.

Source: *Gatwick Area Conservation Campaign*

County will resist night fly pressure

By Staff Reporter

Pressure for more night flights from Gatwick is growing - and so are demands for a second runway, the county council has heard.

Councillor Bill Buck said people living in Crawley, Horsham, East Grinstead and Haywards Heath already suffered from night noise nuisance.

'We should be passing resolutions here declaring that night flights from Gatwick are just not acceptable,' he urged.

Councillor Alf Pegler said moves were again afoot to look at a second runway at Gatwick. 'We always have to be on the ball to take action against such proposals,' he said and added he was worried about proposals which could result from the British Airways-British Caledonian merger.

Councillor Ken Dunn shared concern about night flights. 'Pressure has been mounting among the aircraft operators during the last nine to 12 months to permit a greater throughput of night flights at Gatwick,' he warned.

Some operators had never concealed their conviction that a second runway would not be denied. The argument was that once the airport had achieved the 25 million passenger capacity, then pressure for a second runway would be inescapable.

A great level of vigilance was needed if Gatwick was not to become another blot on the landscape like Heathrow Councillor Arthur Barnett, chairman of the county planning committee, said it would obviously look at the implications of the merger.

A Government consultation paper was expected, and they expected extra pressure for more night flying. 'We will resist it,' Councillor Barnett promised.

Source: *Crawley Observer,* 29 July 1987

Figure 36

Pollution

Apart from the noise which they generate, airports are often associated with other forms of environmental damage. Visitors may have seen the vapour and smoke trails left in the wake of departing aircraft, or have smelt kerosene in the air, or noticed traces of oil on standing water. These manifestations of large airports like Gatwick certainly constitute a change to the natural environment, but do they represent pollution? Is the evidence merely anecdotal, or have scientific studies revealed levels of pollution at Gatwick which are dangerous?

Several studies have been carried out into air pollution at Gatwick. One of the earliest was undertaken by the government's Warren Spring Laboratory (WSL). According to this 1973 study, the airport was a minor contributor to local pollution, emitting the same amount of smoke as a small town of about 2000 people. However, the study was not based on field measurement at Gatwick but on extrapolated data collected at Heathrow. Today there are three times as many aircraft movements per day as in 1973.

The same laboratory was commissioned to carry out another study of air pollution around Gatwick in 1978. Field measurements were made to determine the levels of various atmospheric pollutants including methane and non-methane hydrocarbons, carbon monoxide, nitrogen oxides, smoke and airborne lead.

Carbon monoxide is produced as a result of the incomplete combustion of fuel. Hydrocarbons, the main contributor to the typical odour around Gatwick, are produced in the same way. Aircraft engines are designed for peak efficiency at high power, and are less efficient when idling and at low power such as when the plane is taxiing. This is when the production of these pollutants is greatest. The Warren Spring Laboratory survey measured hourly means for the main pollutants over a five-month period. The results comparing Gatwick with other sites can be seen in Figure 37.

a) Write a brief summary of findings of the WSL study. Would you agree that there is a problem?
b) How might these findings be interpreted by a cynical local journalist who is an active member of the Green Party and a keen rambler? How valid are the comparisons made? Write this journalist's next leader article!

A paper examining air pollution was prepared for the Gatwick Airport Consultative Committee in 1982. It considered all the studies undertaken specifically of Gatwick, as well as work done around other major world airports. The general conclusions were that:

1 airports in general do not have a significant adverse impact on local air quality. Indeed, their air quality is often superior to that in neighbouring urban areas;
2 road vehicles and heating plants are often more significant contributors to pollution at an airport than the aircraft themselves.

Comparison of Gaseous Pollutant Concentrations Measured at Gatwick Airport (Fixed Site) with Other Sites in the UK

Pollutant	Percentiles of Distribution of Hourly Means*		
	50th	90th	99th
Total Hydrocarbons (ppm)			
Gatwick	2.6	3.5	5.1
London Off-Street	3.0	6.0	10.0
London Kerbside	data not available		
Carbon Monoxide (ppm)			
Gatwick	0.2	0.7	1.6
London Off-Street	2.4	4.8	7.9
London Kerbside	3.9	8.5	14.1
Nitrogen Monoxide (pphm)			
Gatwick	0.9	3.4	13.3
London Off-Street	2.9	8.4	17.5
London Kerbside	5.5	22.8	54.0
Nitrogen Dioxide (pphm)			
Gatwick	2.0	3.5	5.1
London Off-Street	2.8	6.7	12.8
London Kerbside	similar to off-street site		

* e.g. a 90th percentile of 0.7 ppm means that 90% of all hourly means measured less than 0.7 ppm.

ppm = parts per million

pphm = parts per hundred million

Comparison of Lead Concentrations Measured at Gatwick Airport (Fixed Site) with Other Sites in the UK

Site	Lead Concentrations (ngm^{-3})
Gatwick	430*
Five Sites Mean 1976/77	510'
Five Sites Mean 1977/78	380'
US Standard	1500**

* Average over the period February-June 1979.

' Mean annual average of sites in Stoke, Leeds (2 sites), Belfast and Coventry

** Applies to an average over a calendar quarter.

ngm = nanogrammes per cubic metre.

Comparison of Monthly Mean Smoke Concentrations at Gatwick Airport (Fixed Site) with some Other Sites in the UK

Month*	Monthly Mean Smoke (µgm)		
	Gatwick Fixed Site	Crawley	20 Site Average'
February	22	21	34
March	17	–	24
April	20	8	19
May	18	11	15
June	26	4	15
Mean	21	11	21

* 1979 data for Gatwick and Crawley (Feb), otherwise 1978.

' Sulphate in Particulate Survey

µgm = microgrammes per cubic metre.

Source: BAA evidence to the 1979 Public Inquiry into the Second Terminal at Gatwick

Figure 37 Air pollution at Gatwick

It was suggested that existing and projected levels of pollutants at Gatwick easily met US pollution standards, and that these themselves are well below the concentrations known to damage health. The implication of this was that although the smell of the airport caused an occasional nuisance to some local residents, it posed no health hazard.

Despite these findings, local residents and interest groups continue to complain of smells and to be concerned about damage to health. Of equal concern, however, are the possible effects of pollution at Gatwick on vegetation. A few years ago Ian Ockenden, a local market-gardener who lives near Horsham, noticed that his output of organically-grown fruit and vegetables was declining. At the same time, soil acidity was increasing, despite heavy liming. His observations suggested that some beech trees were thinning at the crown and showing signs of becoming diseased. Many oak trees too began to suffer from dieback at the crown (see Figure 38). Ian noticed that some of the worst affected areas of tree damage appeared to be in areas overflown by aircraft as they climbed away from Gatwick Airport. Was there, he wondered, a connection between the increase in air traffic at Gatwick and the onset of tree damage?

These observations needed investigating further if they were to be substantiated. Consequently the *Horsham Oak Tree Survey* was devised by Dave Seabrook of Horsham Friends of the Earth and Paul Howlands of the West Sussex Green Party in the spring of 1986. An area of Sussex and Surrey countryside measuring 972 km^2 was selected and sub-divided into 108 squares, each with an area of 9 km^2. The object was to survey the health of mature oak trees with trunks at least 0.6 metres (2 feet) in diameter. Steps were taken not to select trees damaged or under stress from known causes, such as those near roads or whose roots were covered by tarmac.

The survey was undertaken by about 50 members of the public, some of whom were members of local Friends of the Earth groups or the Green Party. They were given basic information about the symptoms of tree damage and crown dieback. Each observer selected a public footpath within their designated 9 km^2 square, and recorded the health of the first 20 oak trees encountered along the path that satisfied the selection conditions. Each tree was recorded as being healthy, slightly or severely damaged, as appropriate. The survey was undertaken between May and June 1986, and covered over 2000 trees in the countryside surrounding Gatwick airport. An assessment was made of overall tree condition in each square, and a grade given. The results were then mapped (see Figure 39).

Figure 38 *Crown dieback in oak trees around Gatwick*

Phases of tree decline due to ageing and stress

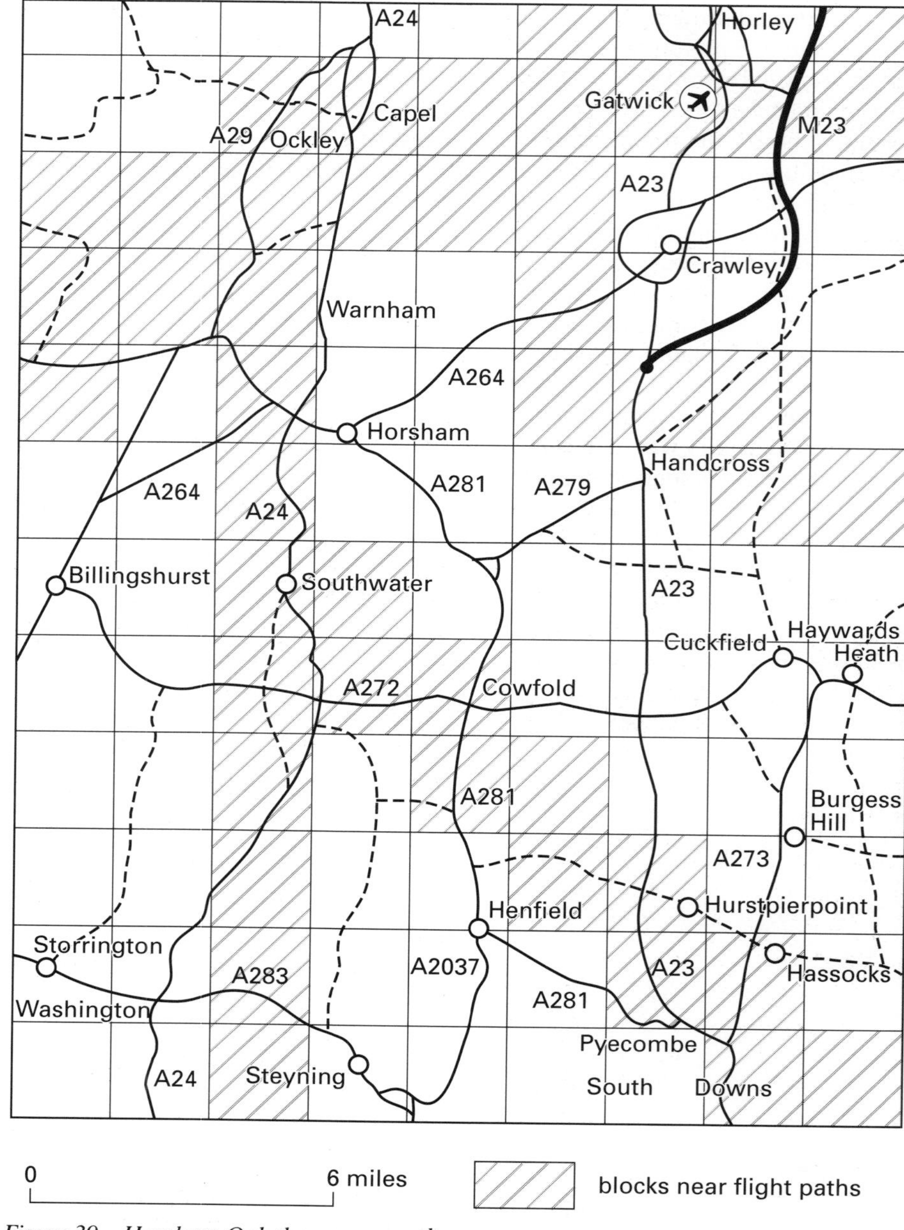

Figure 39 Horsham Oak the survey results

a) Calculate the total area for each of the four categories of oak tree health. What percentage of the study area does each represent?
b) Is there any correlation between Gatwick flight paths and oak tree damage as shown on the two maps?
c) Is the evidence strong enough to prove a link?
d) How could the research design be improved if the data collection were to be repeated?

When the whole study area is included, the average amount of damage for each square was 47% of the area. When the squares under Gatwick's flight paths are examined more closely however, the average damage increases to 57%, compared with just 40% for those squares away from the flight paths. There would then appear to be some evidence to link Gatwick's flight paths and the areas with the worst oak tree damage.

e) If oak trees are being damaged as this survey appears to show, what knock-on effects might this have on the ecology of the area?
f) In groups, try to suggest ideas which might minimise this sort of oak tree damage in the future.

Other local impacts of development at Gatwick

It has already been seen that Gatwick Airport is a major employer. This has important consequences for the local area. As the airport has grown, so the demand for housing in the area has increased. Crawley, Horsham and Haywards Heath have already expanded considerably to house people whose work is either at, or associated with, the airport. Vacant land which can be

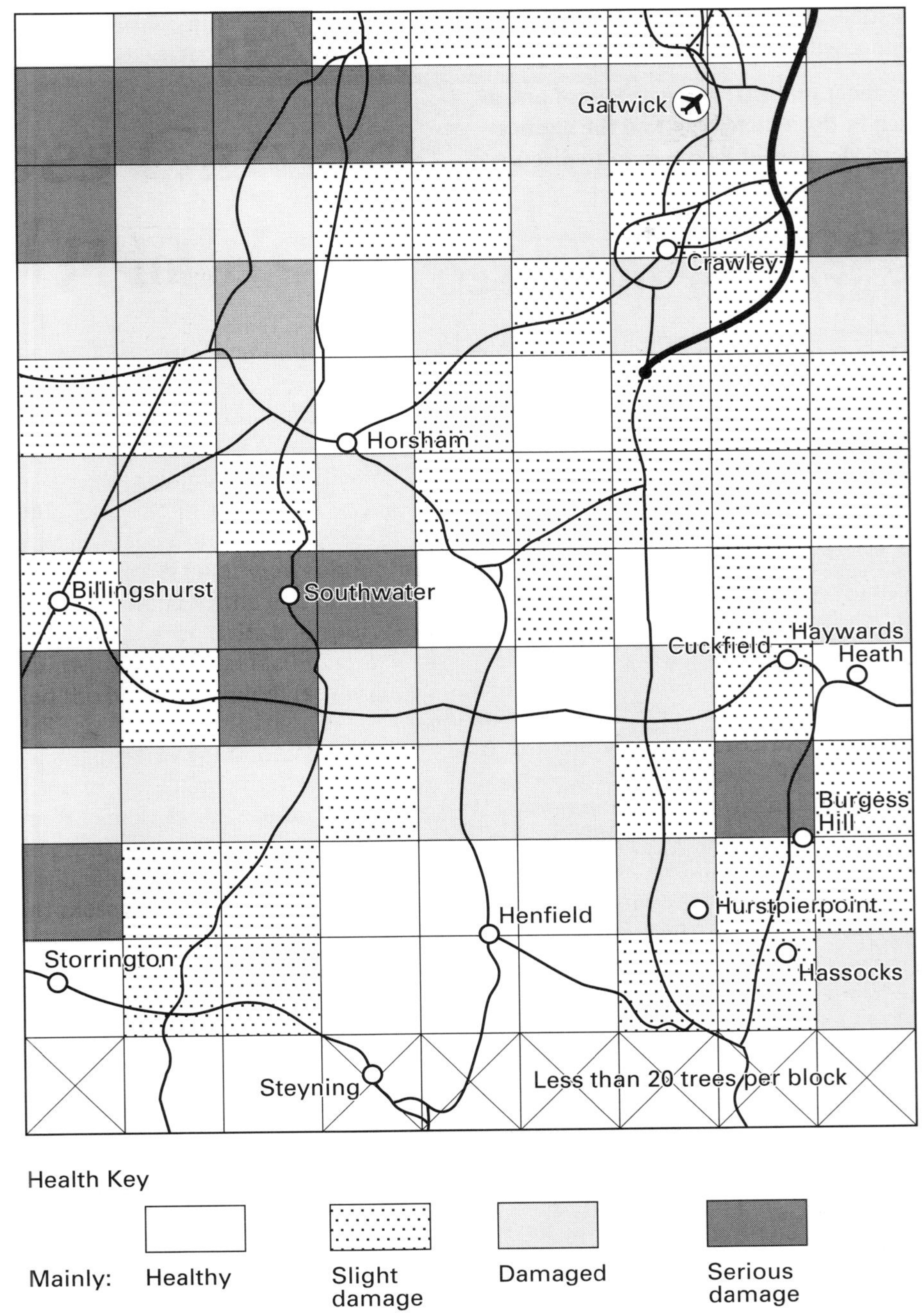

used for housing is difficult to find unless farmland is built upon, and this means changing the character of the area still further.

Continued growth at Gatwick could lead to increases in the price of land and houses. There are also Green Belt restrictions to be adhered to. Local councils have generally argued that the effects of further airport expansion would have adverse consequences for their areas. Additional people would require more shops, schools, hospitals and other amenities as well as roads and associated infrastructure.

At the Second Terminal Public Inquiry in 1979, evidence was given that the residents of towns near Gatwick like Horsham, Haywards Heath, East Grinstead and Burgess Hill were opposed to the planned development because their local physical and social infrastructure was already inadequate as a consequence of the rapid growth they had experienced in meeting the demand for housing resulting from Gatwick's expansion.

Countryside campaigners and amenity groups were likewise opposed to Gatwick's expansion, arguing that the countryside's character and beauty, as well as that of the villages near Gatwick, were being destroyed. The National Farmers Union also maintained that increased urbanisation would lead to a loss of agricultural land and also a loss of productivity. The area's dairy industry was particularly susceptible to vandalism, and the heavy clay soils prevented farmers from growing cereals as an alternative. There was also concern about the possible effects of urbanisation on the catchments of the rivers Mole, Arun and Adur, and the attendant risk of exacerbating flooding.

a) Is the viewpoint expressed by Aileen Walker in Figure 40 one that you have any sympathy with?

b) What would your reaction be if, as a resident of one of the villages mentioned in the article, you had the chance of a meeting with your Member of Parliament to discuss the possibility of further growth at Gatwick?

Air traffic in Britain continues to increase as we have seen. Some extra capacity will have to be provided before long to meet this demand. One possibility is that Gatwick will be expanded by the addition of a second runway. You will be able to explore this option in the next exercise.

Village fears over increased pressure for new homes

By A J McIlroy

From the Tudor porch of St Mary's Church, Billingshurst, Mrs Aileen Walker pointed to a row of terraced houses nearing completion 150 yards away. 'That used to be our unrivalled view of the Sussex Downs. For as far as the eye could see, it was breathtaking. I have treasured it for over 30 years.

'Now we have this long row of modern houses with high roofs and yet another aspect from our church has been changed by so-called progress.'

Billingshurst is among villages deeply troubled over the implications of the call last week by the Environment Secretary, Mr Nicholas Ridley, for an increase of more than 10 per cent on the number of new homes the county planning authority was intending to allow in West Sussex by the mid 1990s.

Local amenity groups backed by the Council for the Protection of Rural England (CPRE) fear that, by sanctioning 37,900 new homes - 3,600 more than the county planners wanted - Mr Ridley 'is inevitably advancing the erosion of a way of life as hamlets become villages and villages become towns.'

Area CPRE officials cite Billingshurst, in the Horsham district planning area, as a 'classic' example of a Sussex village already damaged by development and still under threat because of the demand for homes generated by London commuters, the expansion of Gatwick Airport and the growth of high-technology industries in the Gatwick-Crawley area.

Renewed fears

Officials expressed concern for Southwater, Pulborough, Warnham, Steyning, Broadbridge Heath and other villages.

There are renewed fears for Field Place, the home of the poet Shelley in Broadbridge Heath, even though one application to build there has been turned down by the Secretary of State.

'We are supposed to be relieved because the county planners were expecting Mr Ridley to sanction even more new building than he has,' said Mr Bob Stone, chairman of the Billingshurst Society campaigning against excessive development.

'We are not relieved. The bitter reality for villages in Sussex and Surrey - and wherever the pressure for development exists - is that once a development is allowed, it is the foot-in-the-door that invariably means more to come.

'We are not suggesting that the developers building the houses in front of the church and blocking the view of the Downs are breaking the law. They are not.

'Doing our best'

'Planning approval was given 13 years ago and they are quite entitled to build there. What we say is that this is an example of how a beautiful English village can be changed by a remorseless and steady development growth.'

A spokesman for Horsham District Planning Authority said: 'We are doing our best to protect Billingshurst and all our other villages.'

Source: *Daily Telegraph*, 17 August 1987

Figure 40 Field Place, home of the poet Shelley in Broadbridge Heath

EXERCISE 6

Should a second runway be built at Gatwick?

Every summer, the news media report flight delays, overcrowded airport terminals and air traffic control difficulties. These are the physical manifestations of a system under stress. Currently, the capacities of air terminals, runways and even some air traffic control systems are close to being exceeded. Even some air lanes have become dangerously busy.

As air traffic continues to increase, the problems will worsen. Already a variety of concerned groups are calling for action. Some Members of Parliament, airport planning professionals and the travelling public have made their views known. Lobby groups like SCREAM (Sufferers Campaign to Resolve the European Aviation Mess) have become established. They believe that twice as many passengers will be flying around Europe by the year 2000. By then, three-quarters of the terminals and two-thirds of the runways are forecast to be saturated. They argue that if the system is to have any chance of meeting expected demand then planning must begin now.

SCREAM is supported by the Air Transport Users Committee (AUC), which has long been demanding increased runway capacity for south-east England. Another of SCREAM's backers is the International Foundation of Airline Passenger Associations (IFAPA), who have long been lobbying for infrastructure expansion on a European scale. SCREAM is campaigning to achieve:

1 a liberal, expanding, uncongested air transport system
2 better use of airports and airways today and expansion for tomorrow
3 a more realistic balance of environmental and transportation needs to ensure busy, quiet and passenger-friendly airports
4 a European Master Plan for airport and airspace development which includes pan-continental airspace management.

Whether you agree with these aims or not, list some of the considerations you think they ignore. What interest groups and individuals would you expect not to endorse SCREAM's objectives?

Whatever the disagreement over the precise figures, it seems clear that all the airports in the south-east of England are likely to grow between now and the end of the century (Figure 41). It was to accommodate some of the projected increase that some specific recommendations were made by the House of Commons Select Committee on Transport. As we saw earlier, the 1989 report on Air Traffic Control Safety recommended that:

> in spite of the many difficulties and obstacles and the inevitable controversy, all efforts should be made to provide a second main runway at Gatwick.

Anger at second Gatwick runway

By Nicholas Schoon

THE SELECT Committee's unequivocal recommendation for a second runway at Gatwick brought an instant and angry rejection from councils and MPs.

The Government will find it a highly unpalatable suggestion. The world's second busiest international airport is surrounded by a Tory-voting heartland, and any suggestion that the proposal should be taken up will create enormous anger.

Building the new runway in order to meet a doubling in flights over the next 16 years would require Parliament to set aside a legal agreement made between the BAA and West Sussex County Council in 1979 that no application would be made to build a second runway for the next 40 years.

A new runway, parallel to the existing one, would be the simplest solution but it would cause further massive disruption in a once rural area which is already struggling with one of the highest growth rates in the country.

Peter Bryant, director of planning for West Sussex, said: "The MPs are flying in the face of reality, and I think they're being wholly irresponsible. They can't have examined the practical issues so their comments must be of limited value." Michael Sanders, chief executive of Crawley Borough Council which includes the airport, said: "We're appalled. We've always been supportive of Gatwick, but we could not cope with the extra growth a second runway would cause."

Both councils and BAA pointed out the problems in finding space for a second runway following a government decision 18 years ago to stop reserving land for one. Immediately to the south is the new town of Crawley — which has mushroomed from 8,000 people to almost 90,000 in the past 40 years — and the busy A217 road.

All of Gatwick's apron space and both terminals are north of the existing runway, so aircraft using a new southern runway would have to cross the existing one, severely restricting its capacity. To the north, the best location for another runway has been covered by the new £200m North Terminal.

So a new runway would have to be built further north, obliterating the historic village of Charlwood with its 800 homes, and causing extreme noise problems for Horley's 18,000 inhabitants.

Source: *Independent*, 22nd March 1989

Figure 41

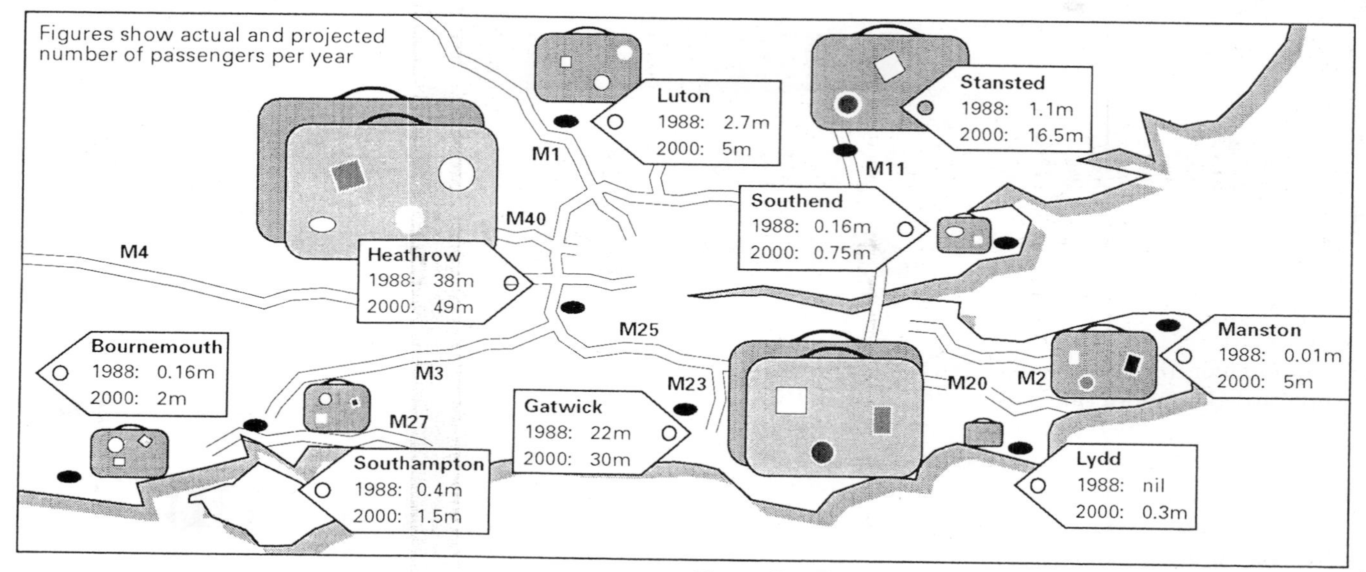

Figure 42 Estimated increase in traffic at airports in south-east England, 1988–2000

Not surprisingly, the reaction to this suggestion was swift and strong (see Figure 42), particularly from those living near Gatwick. Apart from the intrusion that such a development would probably cause, much of the anger generated was because the construction of a second runway at Gatwick would break a 1979 agreement between the British Airports Authority and West Sussex County Council stipulating that one should not be built for 40 years (that is, before 2019). This agreement was confirmed in the government's 1985 White Paper on *Airports Policy* (Cmnd 9542).

The British Airports Authority plc and Civil Aviation Authority agree that more runway capacity will be required in south-east England between 2000 and 2005. They agree also that once a decision is taken about where it is to be located, it will be about 10 years before the extra capacity comes into operational use. Given these considerations, an initial decision needs to be taken soon about where to locate the extra runway.

There are a number of options available:

1 a second runway at Stansted Airport in Essex
2 major expansion at RAF Manston in Kent, where already some charter operations are based at new facilities which operate as Kent International Airport (KIA)
3 combined expansion at several south-east airports including Luton, Bournemouth, Southend, Southampton and Lydd
4 a second runway at Gatwick Airport in Sussex
5 additional facilities including a new runway and an extra terminal at London Heathrow.

a) Divide the class into five groups. Each group should choose one of the above options and discuss it fully. Try to take account of environmental, economic, social and political considerations. List the advantages and disadvantages associated with the option, and try to reach a consensus about its relative merits. You may wish to use extra resources like maps.

b) When the class has reassembled, a spokesperson from each group should summarise the group's views, so that the whole class can get a flavour of the problems and possibilities associated with each option.

c) Which option do you find the most attractive? Write a newspaper article reporting the discussions which have occurred along with any personal conclusions.

The fourth option listed, as has already been pointed out, is the preferred option of the House of Commons Transport Select Committee. But would such a proposal be passed at the public inquiry which would inevitably be called to investigate the planning application?

You now have an opportunity to take part in the sort of public inquiry which would be called if Gatwick Airport Limited (the company which operates Gatwick on behalf of the British Airports Authority plc) made a planning application for a second runway at the airport. This role-play will enable you to use the ideas and knowledge you have acquired in studying this unit to argue from a particular standpoint for or against a second runway at Gatwick.

The public inquiry is designed for nine people, but extra roles can be added. Eight individuals are identified on role cards (see Figure 43) along with the organisations they represent. Each individual will have a particular viewpoint relating to the second runway proposal. The main players in the public inquiry *may* adopt these positions:

In favour of a second runway:	*Against a second runway:*
British Airports Authority plc	Charlwood Parish Council
Crawley Borough Council	Friends of the Earth
Gatwick Area Hoteliers	Country Landowners Association
Airline Pilots Association	National Farmers Union

In addition there should be a neutral chairperson who will control the meeting.

Jenny Barrington

President of the Sussex Branch of the National Farmers Union.

Your organisation represents the interests of local farmers. There is a mixture of smallholders with specialised market gardening operations, and large-scale farmers with huge estates involved in mixed arable and pastoral production. Increasing numbers of your members are switching to organic production involving only natural fertilisers. Others are taking part in the government's 'set-aside' scheme, which has seen part of their land taken out of agricultural production. Most of the land has been in the families of its present owners for generations, thus the links between people and their land are very strong.

Sir Norman Scott

President of the Gatwick Area Hoteliers Association.

The hotels in the Gatwick area owe their livelihood to the airport. They provide overnight accommodation for people travelling from other parts of Britain catching early flights, as well as being a base for tourists visiting this country. They also have large conference facilities and play host to important international conferences. The owners of the hotels have invested substantial sums of money in providing excellent facilities at Gatwick, and their use must be maximised if the hotels are to continue to be profitable. The hotels are important local employers.

Mr Hugh 'Tommy' Whyte

Southern Area Director of the Country Landowners Association.

The area around Gatwick is a rural one, and much of the land is divided into large estates owned by members of the Country Landowners Association. Your organisation aims to protect their interests and is concerned that huge areas owned by CLA members may be lost if the second runway goes ahead. While the land may increase in value, its use would inevitably change and CLA members would have to develop new interests in leisure-associated industries rather than those more traditionally associated with agriculture and the rural economy. Your organisation is a rather conservative one and dislikes change.

Beatrice O'Shaughnessy

Public Relations Officer for Crawley Borough Council's Planning Department

Crawley has derived a considerable amount of prosperity in the past from the proximity of Gatwick. Much employment depends on the continued existence of the airport, and these factors are likely to influence the Borough Council's stance on the second runway issue. The physical infrastructure of the area has also benefited as a result of the airport. Road and rail links with London and the rest of the country are arguably better than they would have been without the airport's presence.

Lt Colonel Marcus D'Arcy KBE

Secretary of Charlwood Parish Council and resident.

As a longstanding resident you will want to represent the views of local householders and residents groups, as well as those of Charlwood Parish Council. Charlwood is a small village located W N W of the airport, and fairly close to flight paths and the airport perimeter. It is overflown regularly by planes and there is considerable disturbance to the residents of this mainly rural location. Your objections centre on the noise, extra traffic and loss of Green Belt land that would inevitably result from permission being given for a second runway.

Justice Elizabeth Duke Q.C.

Inquiry Chairperson

You have been asked by the Department of the Environment to Chair the Public Inquiry into the Second Gatwick Runway Proposal. After nine years working your way up through the ranks of the judiciary, you have previously chaired two Inquiries. The first was concerned with a motorway extension in South Wales and the second focused on the development of a light railway system in a city in north-east England. You were a member of the Heathrow Airport Consultative Committee. You are also a past president of the Noise Abatement Society.

Captain Mike Johnston

Secretary of the British Airline Pilots Association (BALPA).

BALPA members enjoy flying into Gatwick airport. The facilities there are good, and the skies relatively uncrowded compared with those over Heathrow. The surrounding environment is also much more pleasant, and a number of pilots have their homes here. They favour the idea of expansion at Gatwick as this means that more of them will be able to adopt Gatwick as their 'home' airport. Nearby facilities such as schools and hospitals are also good, and the pace of life more relaxing than in the Heathrow area.

Ms Susie Baxter

Sussex Area Liaison Officer for Friends of the Earth

Environmental considerations are foremost among those you will be seeking to draw to the attention of the Inquiry. You actually took part in the Horsham Friends of the Earth Oak Tree Survey, and were very alarmed by its findings. As a primary school teacher and a mother of three you are particularly concerned that the airport will spoil the rural character of your home village of Rusper.

Mr Alan MacIntyre FCIT

Director - Planning Division British Airports Authority plc.

Your position is that the British Airports Authority must ensure that adequate terminal and runway capacity is provided to meet anticipated demand. You see the new North Terminal at Gatwick as providing extra capacity which could be supplemented by the addition of a second runway. This would also help to take the strain off Heathrow Airport.

Figure 43 Role cards for public inquiry

a) Each member of the group should take on one of the roles identified in Figure 43. Think about the role carefully and research it where necessary. Your aim is to create a full and convincing case either for or against the proposal for a second runway, around the personality you create. Make your own decision about the particular viewpoint likely to be taken by the group you represent.

b) The chairperson will ask speakers to give a 5-minute (maximum) presentation to the inquiry. Speakers for and against the proposal will be called alternately. Each presentation will be followed by a question session.

c) At the end of the inquiry the chairperson will ask one member of each group to summarise the evidence presented. A vote should be taken as to whether a second runway should be built at Gatwick, based on the strength of the arguments presented during the inquiry.

d) After the public inquiry, compile a newspaper article setting out the main highlights of the meeting, and the inquiry's findings. While both sides of the case must be reported, you may reveal your own (and your newspaper's) bias. Your article should be supported by an appropriate headline, and include a photograph, sketch or cartoon.

Whatever happens, it is clear that some additional runway capacity needs to be provided in the south-east of England. The figures suggest that it will certainly be needed some time after the turn of the century, and with a 7–10 year gap almost inevitable between the go-ahead being given and the opening of the runway, a decision must be made soon.

Review and Cue

Whatever decisions are taken in the south-east of England, the debate about airport location and growth looks likely to resurface in other parts of Britain in future years. This is partly due to increasing environmental awareness on the part of the general public. It is also due to the inexorable growth in air transport across the whole of Europe.

The cost of air travel inside Britain as well as within continental Europe is becoming increasingly competitive with rail travel. Moreover, political pressure seems likely to achieve a further deregulation of prices. This ought to make air travel in Europe as cheap, kilometre for kilometre, as it is now in the USA. There will be an obvious 'knock-on' effect on the number of passengers wishing to take advantage of the convenience and speed of travelling by air.

Make a comparison between the cost of air travel inside Europe with that inside the USA. Find out the standard single fare by looking in the ABC Air Guide, available in reference libraries, or consulting a travel agent. Use this information to calculate the cost per kilometre of a standard economy single fare on each of the following journeys:

i Paris to Athens
ii Helsinki to Lisbon
iii London to Edinburgh
iv Boston to Washington
v Chicago to Miami
vi Los Angeles to Houston

What conclusions do you draw between pricing policy in the USA and that in Europe?

The popularity of air travel in the United Kingdom for journeys overseas as well as internally has already been referred to. Many provincial airports are undergoing massive growth to try to take advantage of the economic benefits which expansion may bring to them, and the areas in which they are located. At Luton, Manchester and Birmingham plans are already underway, and the opening of London City Airport in Docklands has brought about new scheduled services to a variety of destinations.

The demand for new runway capacity in the south-east may seem, from current trends, inevitable. However, there are other options available which may allow the inevitable decision to be postponed. There are perhaps three main alternatives (see Figure 44). These involve operating the existing system in different and more efficient ways:

1 *Mixed-mode runway operation* At present the two runways at Heathrow Airport are used in 'segregated mode'. This means that at any one time, each runway is used exclusively for landings or take-offs. Instead, each runway could be used for both landings and take-offs. This move would create an extra 50 000 slots for aircraft at Heathrow, and would represent a more efficient use of runway capacity, and is already common practice in the USA. Here, it would require changes in the air traffic control systems.

Figure 44 Increasing the efficiency of existing runway capacity

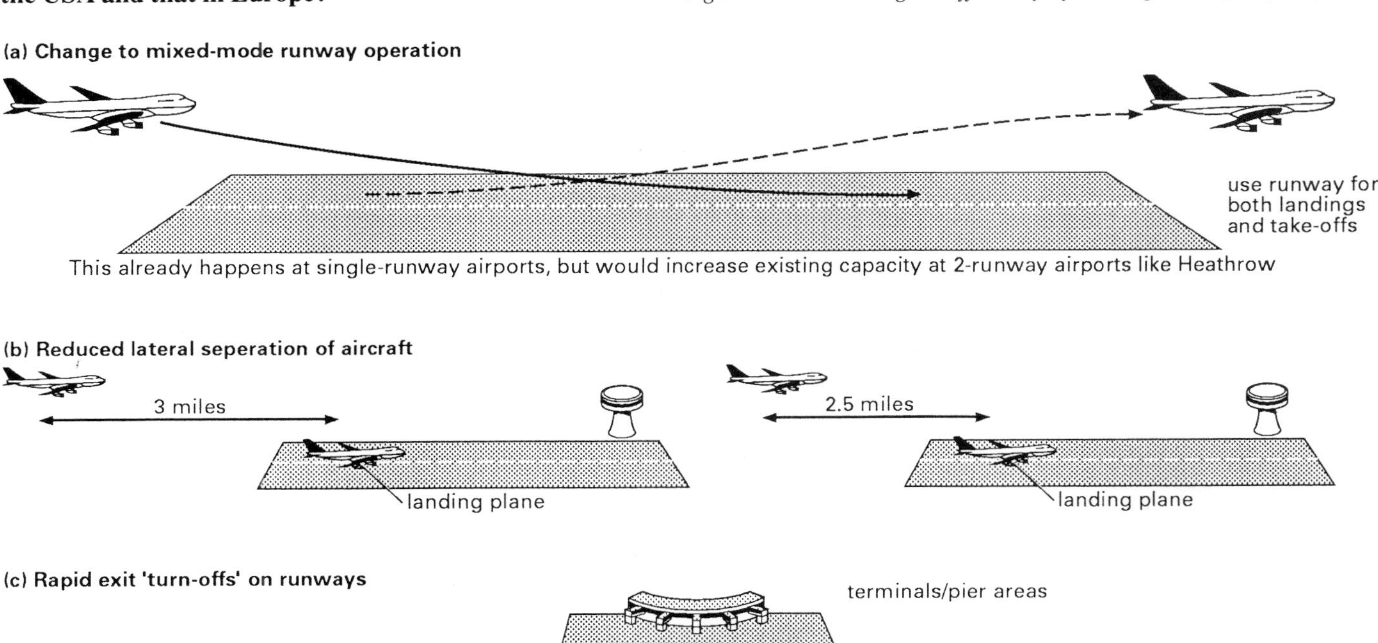

2 *Lateral separation of aircraft* The Civil Aviation Authority's regulations demand that there must be a 3-mile (4.83 kilometre) lateral separation between all landing aircraft. This means that when a plane lands, the aircraft waiting to land behind it must be at least 3 miles from the runway. In the USA a lateral separation of only 2.5 miles has been operated quite safely. This measure would create an extra 5000 landing slots a year at Heathrow.

3 *Rapid exit 'turn-off'* Once an aircraft has actually landed, all the time it is on the runway it is blocking other planes which are ready to land. The quicker a plane can get off the runway after landing, therefore, the more time is provided for other aircraft. The long runways provided at large international airports today are not required for landing purposes by many smaller aircraft. If 'turn-offs' were provided for such planes, enabling them to turn off the runway earlier than at present, the runway could be used more efficiently. Such a move would produce about 2500 extra slots per year.

Would you be in favour of introducing any of these 'efficiency' measures? What drawbacks might they have? Are they preferable to the expansion of existing airports?

Although efficiency measures might help the situation in the London area, they would probably only postpone the time when a decision on further runway capacity needs to be taken. Unless unforeseen circumstances intervene, it seems that the number of passengers travelling by air is likely to continue to increase. The more widespread concern about the environment is sure to result in growing conflicts between those lobbying for airport growth, local residents and environmental groups.

Opposition to airport growth in Britain has been fairly controlled and generally within the law. However, might the prospect of a second runway in the Sussex countryside at Gatwick be too much for some people to accept? Could it trigger the ferocious opposition that the construction of the new Narita international airport in Tokyo did in the early 1970s? Then, tear-gas and violence (see Figure 45) were used by demonstrators.

Try to research the background to the opposition to the building or expansion of an airport somewhere in the world. You might choose Narita in Tokyo, or a more local example. In your view, is using violence ever justified in the cause of 'protecting' the environment from airport development?

A way may yet be found of solving the problems in the south-east of England without upsetting people. There are locations in which airports could be built where no people live. In the late 1960s for example there was a proposal to construct London's third airport at Maplin Sands on shingle banks in the Thames estuary. It did not succeed on that occasion but it is perhaps not surprising that a similar proposal (Figures 46 and 47) was put forward in March 1990 by a private consortium.

Figure 45 Japanese protesters against airport development

Consortium plans London airport in Thames estuary

by David Black, Transport Correspondent

Plans for a privately-funded London airport to be built on a man-made island in the Thames estuary, will be presented to the Government within the next month. The chosen site is west of the West Shingles Bank in mid-estuary. It is about eight miles south of Foulness, Essex, and a similar distance north of the Isle of Sheppey, Kent.

An international consortium has had London-based architects researching its feasibility for two years. The five main backers - one UK-based and the rest American and Australian investors - believe it will be vital for London, if predicted rises in airline passenger demand to 2005 are to be met.

An airport equivalent in size to Heathrow, built on a 3,000 acre island above sandbanks which at present are covered by less than 7ft of water at low tide, is envisaged. It would be linked by tunnels - each carrying rail and multi-lane roads - to Kent and Essex. It would eventually have three runways, and a capacity of 45 million passenger movements a year.

There are contingency plans to tunnel under areas where there may be environmental objections. A previous attempt to build an airport off Foulness, on land reclaimed from Maplin Sands, failed because objectors claimed it would lead to the destruction of wildlife on the coastal marshes.

According to architects Covell Matthews International, initial costings for the project envisage launch funding of £3.5bn over the first five years. This would build the island and the first of three runways. There would also be terminal facilities for 15 million passenger movements a year. Remaining runways and terminals would be added over five years bringing the total cost to £7.5bn.

Jerry Matthews, managing director of Covell Matthews, said: 'By the year 2005, the CAA, BAA, and the government have all agreed on traffic forecasts of between 105 and 123 million passenger movements a year. Even if you add Heathrow, Gatwick and Stansted, and all their extra runways and new terminals, it still means there will be between 40 and 45 million journeys that can't be made.'

Source: *The Independent*, 9 March 1990

Figure 46

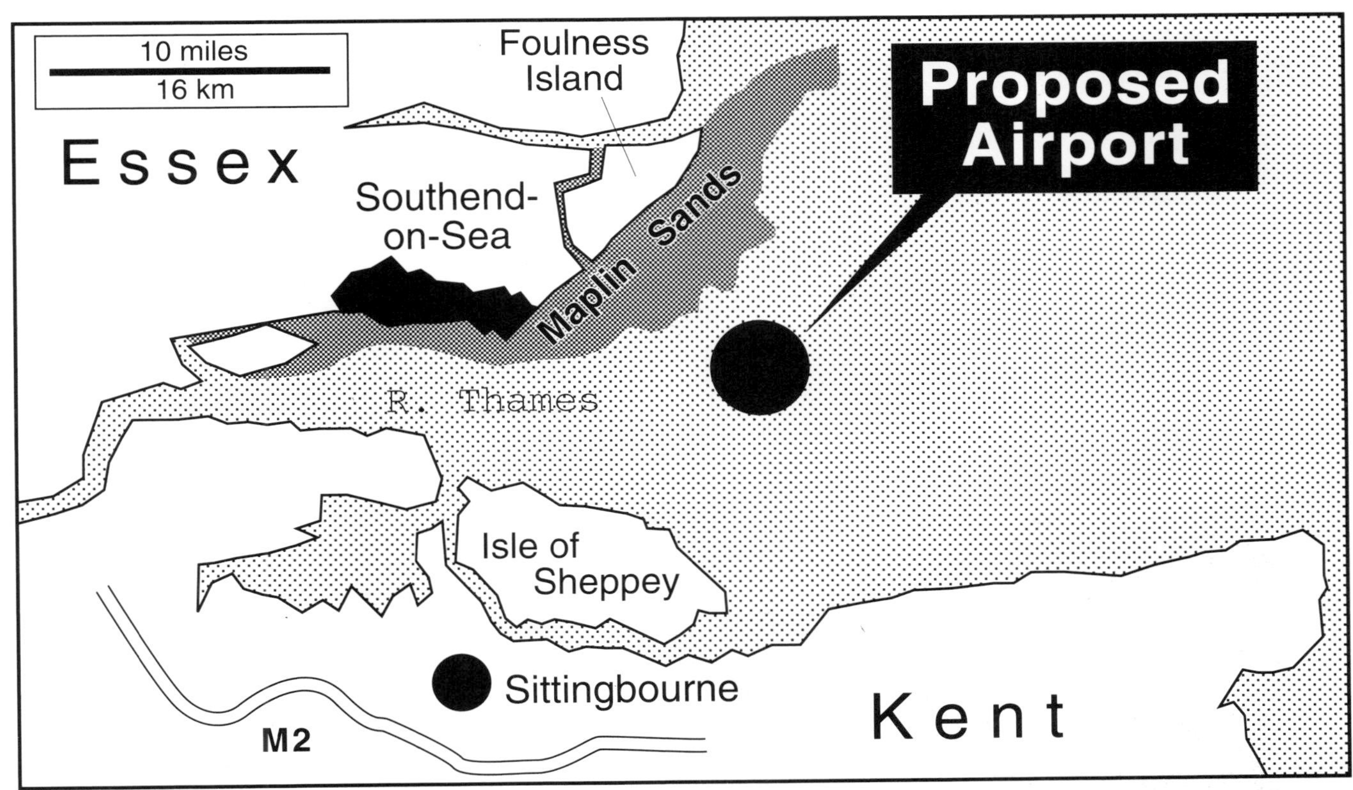

Figure 47 Location of the proposed Thames Estuary airport

Is this proposal, in your view, preferable to the construction of a second runway at Gatwick? What implications does it have?

The timing of the construction of new runway capacity in the London area may be in doubt, but the need for it is not. Controversy is almost inevitable. Large and complex developments like modern airports are bound to have a range of repercussions for the environments in which they are located. Moreover, the situation may be insoluble, in that the volume of traffic may simply expand to fill whatever capacity is provided!

Bibliography

BAA Statement of Case: Gatwick Airport Public Inquiry, (BAA plc, August 1979)

BAA: Into the 1990s: Gatwick Airport Master Plan Report 1983, (BAA plc, 1983)

BAA Airports – Traffic Statistics 1988–89, (BAA, 1989)

BAA: Financial and Operating Information (Supplement to 'Annual Report and Accounts', (BAA plc, 1989)

Bishop, M: 'Crisis at Heathrow' in *Voyager*, February 1990 (published on behalf of British Midland Airways Ltd)

Friends of the Earth: Horsham Oak Tree Survey, (Arun Friends of the Earth 1986)

Gatwick Airport Consultative Committee: *GACC Annual Report 1986*

Gatwick Airport Consultative Committee: Annual Report 1988

Gatwick Airport Ltd: The Gatwick Guide (Priory Publishing Ltd., 1986)

Gatwick Area Conservation Campaign: *Good Neighbour Airports*, 1985

House of Commons Transport Committee (1989): *First Report on Air Traffic Control Safety*

Public Inquiry into the Second Terminal at Gatwick Report (1981)

Varley H: *The Flier's Handbook* (Pan Books, 1978)

White Paper on Airports Policy (Cmnd 9542)